£39-75

Communications and Control Engineering Series

Editors: A. Fettweis · J. L. Massey · M. Thoma

Scientific Fundamentals of Robotics 3

M. Vukobratović
M. Kirćanski

Kinematics and Trajectory Synthesis of Manipulation Robots

With 66 Figures

Springer-Verlag
Berlin Heidelberg New York Tokyo

D. Sc., Ph. D. MIOMIR VUKOBRATOVIĆ, corr. member of
Serbian Academy of Sciences and Arts
Institute »Mihailo Pupin«, Belgrade
Volgina 15, POB 15, Yugoslavia

M. Sc., MANJA KIRĆANSKI
Serbian Academy of Sciences and Arts
Institute »Mihailo Pupin«, Belgrade
Volgina 15, POB 15, Yugoslavia

ISBN 3-540-13071-3 Springer-Verlag Berlin Heidelberg New York Tokyo
ISBN 0-387-13071-3 Springer-Verlag New York Heidelberg Berlin Tokyo

Library of Congress Cataloging in Publication Data.
Vukobratović, Miomir.
Kinematics and trajectories synthesis of manipulation robots.
(Scientific fundamentals of robotics ; 3)
(Communications and control engineering series)
Bibliography: p.
Includes index.
1. Robotics. 2. Machinery, Kinematics of.
I. Kirćanski, M. (Manja) II. Title. III. Series.
IV. Series: Communications and control engineering series.
TJ211.V839 1986   629.8'92   85-27787
ISBN 0-387-13071-3 (U.S.)

Offsetprinting: Mercedes-Druck, Berlin
Bookbinding: Lüderitz & Bauer, Berlin
2161/3020 5 4 3 2 1 0

Preface

A few words about the series "Scientific Fundamentals of Robotics"
should be said on the occasion of publication of the present monograph.
This six-volume series has been conceived so as to allow the readers to
master a contemporary approach to the construction and synthesis of control for manipulation robots. The authors' idea was to show how to use
correct mathematical models of the dynamics of active spatial mechanisms for dynamic analysis of robotic systems, optimal design of their
mechanical parts based on the accepted criteria and imposed constraints,
optimal choice of actuators, synthesis of dynamic control algorithms
and their microcomputer implementation. In authors' oppinion this idea
has been relatively successfully realized within the six-volume monographic series.

Let us remind the readers of the books of this series. Volumes 1 and 2
are devoted to the dynamics and control algorithms of manipulation robots, respectively. They form the first part of the series which has a
certain topic-related autonomy in the domain of the construction and
application of the mathematical models of robotic mechanisms' dynamics.
Published three years ago, the first part has provided a foundation for
both the expansion of the previously achieved results and the generation of new, complementary results which have found their place in the
following volumes. This is now the second part of the series, consisting of four volumes, has been created. While Volumes 5 and 6 deal with
the problems of nonadaptive and adaptive control and computer-aided
design of robots, book 4 presents new results attained in the domain of
numeric-symbolic dynamic modelling of robotic systems in real time.

This volume mostly deals with manipulator kinematics. Most efficient
methods for evaluating the manipulator kinematic models and the motion
generation algorithms for nonredundant and redundant manipulators are
considered. The complete material of this monograph is divided into six
chapters.

In Chapter 1 main mathematical relations describing the manipulator end-effector position and orientation are presented. Two methods for kinematic modelling are outlined. The first method makes use of Rodrigues formula describing a finite rotation of a rigid body about a fixed axis, while the second approach uses the homogeneous transformations with Denavit-Hartenberg kinematic notation. The specification of the manipulator end-effector position by means of spherical and cylindrical coordinates, together with Cartesian coordinates is described in this chapter. Jacobian matrices relating between linear and angular velocities of the end-effector and joint coordinates are derived too. Efficient methods are presented for computing these matrices.

In Chapter 2 a method for the automatic, computer-aided generation of a computer program for evaluating the symbolic kinematic model of an arbitrary serial-link manipulator is presented. The computer program obtained contains the minimal number of floating-point multiplications and additions. This approach to kinematic modelling of manipulation robots is preferred mainly because of its automation, which results in the elimination of the errors which could easily occur in the manual derivation of these equations. With regard to the advantages of this automatization, the necessity for numeric evaluation of kinematic models is completely eliminated in the motion synthesis of industrial robots.

In Chapter 3 the inverse kinematic problem is considered. Two main methods for obtaining joint coordinates, given the position and orientation of manipulator end-effector, are discussed. The first refers to analytic methods, where joint angles are expressed as the explicit functions of the external coordinates and the already evaluated joint coordinates. Such a solution is only feasible for kinematically simple manipulators. On the other hand, numeric methods are computationally more complex and consequently more time consuming. Besides, the singular points, where the manipulator becomes degenerate complicate the application of numerical methods in motion generation.

In Chapter 4 methods for manipulator motion generation are reviewed. They provide for the continuity of position, velocity and acceleration, which are so desirable in robot motion. At the same time the functionality of motion is achieved by attaining the positions and orientations required. The methods are simple and appropriate for practical application. They are based on manipulator kinematic equations only, and no dynamic characteristics of the system are taken into consideration.

In Chapter 5 deals with the manipulator motion synthesis where the system is considered as a dynamic system modelled by the complete, non-linear dynamic model of the mechanism and the actuators. The complexity of the solution to the general optimal control problem for such systems, is illustrated. Regarding the practical importance of the energy optimal motion synthesis, which simultaneously provides for smooth, jerkless motion and minimal actuators' strains, a particular attention was paid to the energy optimal motion of nonredundant manipulators. An algorithm for determining the energy optimal velocity distribution, given a manipulator end-effector path, was presented too.

In Chapter 6 the problems connected with redundant manipulator motion synthesis are discussed. The various methods, aimed at resolving the ambiguity of the inverse kinematic problem, which have been proposed to date are presented. The motion synthesis in free work space and obstacle-cluttered environment are discussed separately.

Similarly to the previous volumes of the Series, the present book is also intended for students enrolled in postgraduate robotics courses and for engineers engaged in applied robotics, especially in studying kinematics of robotic systems. The authors also think that this, essentially monographic work may, with no great difficulties, be used as the teaching material for subjects from the field of robotics at graduate studies.

The authors are grateful to Mrs. Patricia Ivanišević for her help in preparing English version of this book. Our special appreciation goes to Miss Vera Ćosić for her careful and excellent typing of the whole text.

September 1985   The authors
Belgrade, Yugoslavia

Contents

X

Chapter 1
Kinematic Equations

1.1. Introduction

In this chapter we will consider some basic kinematic relations describing manipulator motion synthesis. They include the relationship between external (world) coordinates describing manipulator hand motion and joint coordinates (angles or linear displacements). The different approaches to kinematic modelling which have been developed in the last years will be discussed.

The first problem to be considered is that of computing the manipulator tip position with respect to a reference, base coordinate frame, given a vector of joint coordinates. The specification of manipulator hand orientation is of importance in describing manipulation tasks. Therefore, the problem of how to determine hand orientation, given a manipulator configuration in space, will be also described.

The relationship between the linear and angular velocities of the manipulator end-effector and the joint rates is also very important in manipulator control. We will here discuss various types of Jacobian matrices relating between these velocities.

The problem of computing the joint coordinates, given a manipulator position and orientation, will be studied in Chapter 3. The complexity of the inverse kinematic problem, together with various approaches to obtain its solution, will be considered, too.

1.2. Definitions

In this section we will introduce some basic notations and definitions relevant for manipulator kinematics formulation. We will be concerned with the manipulator structure, link, kinematic pair, kinematic chain, the joint coordinate vector and its space, the external coordinate vector and the external coordinate space, direct and inverse kinematic problems and redundancy.

Let us consider the model of a robot mechanism shown in Fig. 1.1. The model consists of n rigid bodies which represent mechanism links. These links are interconnected by revolute or prismatic (sliding) joints, having rotational and translational motion, respectively.

Fig. 1.1. Model of a manipulator with n links and n joints

Manipulator structure

The mechanical structure of the mechanism, represented by an arranged n-tuple $(J_1,...,J_n)$, will be termed as manipulator structure, where for each $i \in N = \{1,...,n\}$, $J_i \in \{R, T\}$. Here R stands for a revolute joint and T for a prismatic one.

For example, the manipulator structure RTTRRR stands for a mechanism with 6 joints, where the second and third joints are sliding, and the remaining joints are rotational.

Link

A link is defined by an arranged set of parameters $C_i(K_i, \mathcal{D}_i)$, where K_i represents a set of kinematic parameters and $\mathcal{D}_i$ a set of dynamic parameters.

The sets K_i and $\mathcal{D}_i$ may be defined in various ways. Precise definition of kinematic parameters for Rodrigues formula approach and Denavit--Hartenberg notation will be presented in Subsections 1.3.1 and 1.3.2.

The set K_i includes a local coordinate frame attached to link i, a set of distance vectors (parameters) describing the link i and unit vectors of joint axes. On the other hand, the set of dynamic parameters D_i involves the mass of the link i and its tensor of inertia $\underline{\underline{J}}_i$. If the unit vectors of the local coordinate frame of link i coincide with the main central axes of inertia, this tensor reduces to three moments of inertia $\underline{\underline{J}}_i = (J_{i1}, J_{i2}, J_{i3})$.

Kinematic pair

A kinematic pair P_{ik} represents a set of 2 adjacent links $\{C_i, C_k\}$ interconnected by a joint in point z_{ik}.

The notion of the class and the subclass of a kinematic pair is introduced depending on the type of joint connection. A *j*th class kinematic pair (j=1,...,5) is defined as a set of 2 adjacent links interconnected by a joint with n = 6-j degrees of freedom. A kinematic pair of *j*th class and *l*th subclass is defined as a pair having r rotational and t prismatic joints in point z_{ik}, where

$$r = \begin{cases} m-l+1, & l \leqslant m+1 \\ 0, & l > m+1 \end{cases}$$

$$t = s-r$$

m denotes the maximum possible number of rotational degrees of freedom in the *j*th class. For example, classes 1, 2 and 3 permit 3 rotations (m=3), class 4 - two, and class 5 only one rotation.

Kinematic chain

A kinematic chain Λ_n is a set of n interconnected kinematic pairs, $\Lambda_n = \{P_{ik}\}$, $i \in N$, $k \in N$.

According to the structure of connections, chains are classified into simple, complex, open and closed.

A chain in which no link C_i enters into more than 2 kinematic pairs is said to be a simple kinematic chain. On the other hand, a complex kinematic chain contains at least one link C_i, $i \in N$ which enters into more than 2 kinematic pairs.

An open kinematic chain possesses at least one link which belongs to one kinematic pair only. If each link enters into at least two kinematic pairs, the chain is said to be closed.

In this book we will consider only simple open kinematic chains.

Joint coordinates

Scalar quantities which determine the relative disposition of the links of the kinematic pair $P_{i,k} = \{C_i, C_k\}$ are reffered to as manipulator joint coordinates q_{ik}^{ℓ}. The superscript $\ell \in \{1,\ldots,s\}$, where $s=6-j$ is the number of degrees of freedom, and j is the class of pair P_{ik}.

For the fifth-class pairs, having a single degree of relative motion between links C_{i-1} and C_i, the joint coordinate is q_i. However, a reference disposition of the links which is considered as initial, i.e. where $q_i = 0$, can be chosen in different ways, depending on the manner in which link coordinate frames are attached to the links. We will describe in Subsection 1.3.1 and 1.3.2, how the initial links dispositions, for revolute and sliding joints, are determined, in the two main kinematic modelling techniques.

For an open, simple kinematic chain with n degrees of freedom, the joint coordinates form an n-dimensional vector q

$$q = [q_1 \ q_2 \ \cdots \ q_n]^T \in R^n.$$

Joint coordinate space

Joint coordinate space is the n-dimensional space $Q \subset R^n$, $Q = \{q: q_{imin} \leqslant q_i \leqslant q_{imax}\}$, where q is the joint coordinate vector, q_{imin} and q_{imax}, $i=1,\ldots,n$ are boundary values of joint coordinate q_i, defined by physical constraints of the manipulator mechanical structure, and n is the number of degrees of freedom. The manipulator location in the work space is uniquely defined, given a vector $q \in Q$ (often referred to as manipulator configuration). Joint coordinate space is also termed as configuration space.

External coordinates

External (or world) coordinates x_{ei}, i=1,...,m describe position and orientation (completely or partially) of the manipulator hand with respect
to some reference coordinate system. The reference system is chosen to
suit a particular application. Most frequently, a fixed coordinate frame
attached to manipulator base is considered as the reference system.
More detailed discussion about the reference system with respect to
which the manipulation task is described, will be presented in Chapter 4.

External coordinates x_{ei} form an n-dimensional vector $x_e = [x_{e1} \cdots x_{em}]^T \in R^m$.

The choice of the external coordinate vector as well as its dimension m,
are highly dependant on the given manipulation task and the manipulator
itself. For practical, industrial manipulators, the case m=6 is the most general case, since it makes specification of any payload position and orientation possible. Therefore, we will consider 6-dimensional external coordinate vector (less number of external coordinates is obtained simply by rejecting some of them). A common partition is that the first three elements
of the external coordinate vector x_e define the position of the end-effector,
while the rest of them define orientation with respect to a reference coordinate system. Thus we will distribute vector into two parts

$$x_e = [x_{eI}^T \ x_{eII}^T]^T \qquad (1.2.1)$$

where $x_{eI} \in R^3$ specifies hand position and $x_{eII} \in R^{m-3}$ - hand orientation.
Usually, position of the manipulator hand is specified by

$$x_{eI} = [x \ y \ z]^T \qquad (1.2.2)$$
$$x_{eI} = [r \ \beta \ \alpha]^T \qquad (1.2.3)$$
$$x_{eI} = [r \ \alpha \ z]^T \qquad (1.2.4)$$

where x, y and z are Cartesian coordinates (Fig. 1.2), r, β and α are spherical
coordinates (Fig. 1.3) and, finally, r, α and z are cylindrical coordinates
(Fig. 1.4). In these figures $\vec{q}_{n1}$, $\vec{q}_{n2}$ and $\vec{q}_{n3}$ denote the unit vectors of
the coordinate frame attached to the link n (last link in the chain).

Cartesian coordinates are of primary importance in industrial practice,
while spherical and cylindrical coordinates are convenient for specific
tasks. Spherical coordinates correspond to a translation r along the z
axis followed by a rotation β about y axis, and then a rotation α about
z axis. Specifying the position of the manipulator tip in cylindrical coordinates, corresponds to a translation r along the x axis, followed by a

rotation α about the z axis, and finally a translation z along z axis.

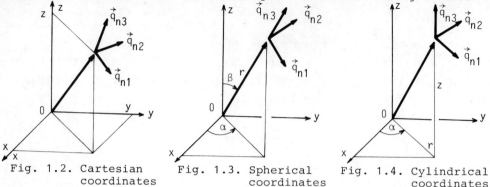

Fig. 1.2. Cartesian coordinates   Fig. 1.3. Spherical coordinates   Fig. 1.4. Cylindrical coordinates

The orientation of the end-effector has been specified in several ways in the references on the subject. Orientation vector x_{eII} may have the form

$$x_{eII} = [\psi \; \theta \; \varphi]^T \qquad (1.2.5)$$

where ψ, θ and φ are Euler angles. Several types of Euler angles have been adopted [1, 3], depending upon the sequence of rotations about the x, y and z axes. Here, we shall consider only yaw, pitch and roll angles (Fig. 1.5), since they seem to be the most appropriate for specifying manipulator hand orientation. Yaw angle corresponds to a rotation ψ about the z axis, pitch corresponds to a rotation about the new y axis, and roll corresponds to a rotation φ about the x axis.

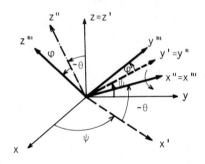

Fig. 1.5. Yaw, pitch and roll angles

Another way to specify hand orientation by means of three magnitudes is the use of Euler parameters [4]

$$x_{eII} = [\lambda_1 \; \lambda_2 \; \lambda_3]^T \qquad (1.2.6)$$

A more detailed discussion on manipulator hand orientation will be given in Section 1.4.

External coordinate space

The external coordinates space is defined as m-dimensional space $X_e \subset R^m$, $X_e = \{x_e: x_e = f(q), q \in Q\}$, where f is a nonlinear, continuous and differentiable vector function, which maps joint coordinates vector $q = [q_1 \cdots q_n]^T \in Q$ into the external coordinates vector $x_e = [x_{e1} \cdots x_{em}]^T$ in a unique way. The space X_e is at the same time a generalized region of reachable manipulator work space.

Direct kinematic problem

Manipulator position and orientation in space, i.e. external coordinates vector is uniquely defined, given a joint coordinates vector $q \in Q$. Solving the equation

$$x_e = f(q) \tag{1.2.7}$$

is known as the direct kinematic problem. This solution differs depending on the type of external coordinates. This will be discussed in Sections 1.3 and 1.4.

Inverse kinematic problem

Determining joint coordinates, given a vector of external coordinates, i.e. solving the equation

$$q = f^{-1}(x_e) \tag{1.2.8}$$

is known as the inverse kinematic problem. This problem is far more complex than the direct kinematic problem, since it is equivalent to obtaining solutions to a set of nonlinear trigonometric equations. This problem will be considered in Chapter 3 in detail.

Redundancy

Depending on the number of degrees of freedom and a given manipulation task, a manipulator can be considered as nonredundant or redundant. A manipulator is regarded as nonredundant if the dimension of the exter-nal coordinates vector (used for discribing the task) is equal to the number of degrees of freedom of the mechanism, i.e. when m=dim x_e = n = = dim q. If n>m is valid, i.e. the number of degrees of freedom is greater than the number of external coordinates, the manipulator is re-dundant with respect to the manipulation task.

In this book we will be predominantly concerned with nonredundant mani-pulators, except for the last chapter, where motion synthesis for re-dundant manipulators will be considered.

1.3. Manipulator Hand Position

In this section we will be concerned with one of the basic problems in manipulator motion synthesis - computing manipulator position in the work space, given a vector of joint coordinates. Position will be spec-ified by Cartesian, spherical and cylindrical coordinates.

Two main kinematic modelling techniques have been used in manipulator control. The first [2, 5-7] was developed as a part of the algorithm for dynamic modelling of open active mechanism. It is based on the use of Rodrigues formula and will be termed Rodrigues formula approach. The second [8], was developed by Denavit and Hartenberg, and is extensively used in robotics today. We will make use of both kinematic modelling techniques in this book, and therefore, we will present them both in the following text.

1.3.1. Rodrigues formula approach

The set of kinematic parameters K_i which are assigned to manipulator link C_i belonging to a simple open kinematic chain, has the form

$$K_i = (\tilde{Q}_i, \tilde{R}_i, \tilde{E}_i)$$

where

$$Q_i = (\vec{q}_{i1}, \vec{q}_{i2}, \vec{q}_{i3})$$ - is a local coordinate frame attached to link i;

$\tilde{R}_i = \{\vec{\tilde{r}}_{ii}, \vec{\tilde{r}}_{i,i+1}\}$ — the set of distance vectors expressed with respect to coordinate system Q_i, assigned to link C_i;

$\tilde{E}_i = \{\vec{\tilde{e}}_i, \vec{\tilde{e}}_{i+1}\}$ — the set of unit vectors of joint axes by which link C_i is connected to links C_{i-1} and C_{i+1}, expressed in coordinate system Q_i.

An example of a manipulator link is shown in Fig. 1.6. Here $\sim$ denotes that the corresponding vector is expressed in the local coordinate system attached to link i.

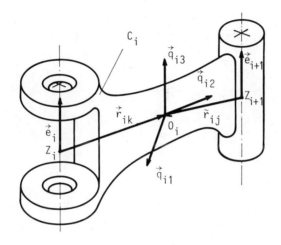

Fig. 1.6. Manipulator link and its kinematic parameters

The local coordinate system is chosen in such a way that its unit vectors $\vec{q}_{i1}$, $\vec{q}_{i2}$ and $\vec{q}_{i3}$ coincide with main axes of inertia of the link, and its origin is positioned into the link mass center. The reasons for such frame assigment are its convenience for dynamic modelling of the mechanism, since in that case, the inertia tensor is always reduced to three main central moments of inertia. However, from the kinematic point of view, this is an arbitrary way of attaching the frame, since in a general case these axes do not coincide with the joint axis or a common normal between joint axes. This results in a more complex kinematic modelling than in Denavit-Hartenberg notation. However, it often happens that main axes of inertia are parallel to joint axis and to the common normal between the joint axes. In that case kinematic analysis in Rodrigues formula approach is also simplified.

Distance vectors which describe the manipulator link C_i are defined in the following way (Fig. 1.6). Vector $\vec{r}_{ii}$ is a vector between center Z_i of the joint i and the origin of coordinate frame Q_i (mass center of link i). Vector $\vec{r}_{i,i+1}$ is a vector from the center Z_{i+1} of joint i+1 to the origin of coordinate frame Q_i.

Unit vector $\vec{e}_i$ corresponds to joint axis about (or along) which joint motion q_i is performed. The notation $\underset{\sim}{\vec{e}}_i$ denotes that it is expressed with respect to the local coordinate frame Q_i of link i. Vector $\underset{\sim}{\vec{e}}_{i+1}$ represents vector of joint axis i+1 given with respect to the same coordinate frame Q_i.

Let us introduce an indicator ξ_i for joint i, denoting whether it is a revolute joint or a sliding one. We will take

$$\xi_i = \begin{cases} 0, & \text{if joint i is a revolute one} \\ 1, & \text{if joint i is a sliding one.} \end{cases}$$

Let us now define strictly the joint coordinate q_i for a revolute joint. Consider the vector $\vec{e}_i$ of the joint axis and the plane Π_i perpendicular to $\vec{e}_i$ at the center of joint i - point Z_i (Fig. 1.7). The center of joint is an arbitrary point at the joint axis. Distance vectors $\vec{r}_{ii}$ and $\vec{r}_{i-1,i}$ are chosen to correspond to the selected joint center Z_i. Denote by $\vec{r}^p_{ii}$ and $\vec{r}^p_{i-1,i}$ the projections of these vectors onto the plane Π_i. The joint coordinate q_i is the angle between vectors $-\vec{r}^p_{i-1,i}$ and $\vec{r}^p_{ii}$ (Fig. 1.7). The situation when these two vectors coincide corresponds

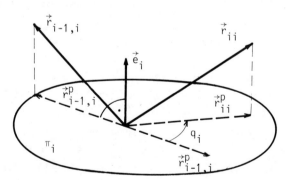

Fig. 1.7. Definition of joint coordinate for a revolute joint

to the zero value of joint coordinate q_i. A more detailed scheme of vectors defining the joint coordinate q_i for a revolute kinematic pair is presented in Fig. 1.8. Here, points O_{i-1} and O_i denote mass centers of links C_{i-1} and C_i, respectively.

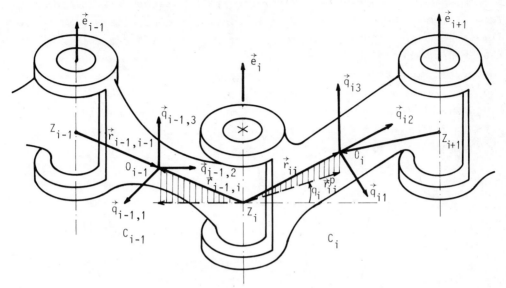

Fig. 1.8. Joint coordinate of a revolute kinematic pair

Let us now consider a special case when either $\vec{r}_{i-1,i}$ or $\vec{r}_{ii}$ is colinear with $\vec{e}_i$. In that case the projection of this vector onto the plane
perpendicular to joint axis $\vec{e}_i$ is not defined, so that the above definition of joint angle can not be applied. Then, it is necessary to introduce an auxilliary vector which is not colinear with $\vec{e}_i$, whose projection will determine joint angle q_i. For example, if $\vec{r}_{ii}$ is colinear
with $\vec{e}_i$ one can adopt vector $\vec{q}_{i1}$ or $\vec{q}_{i2}$ (Fig. 1.8) instead of $\vec{r}_{ii}$, so
that its projection determines the zero value of q_i.

The joint coordinate for a sliding joint is defined in the following
way. The center Z_i of joint i lies on joint axis and is defined by the
initial point of vector $\vec{r}_{i-1,i}$ (Fig. 1.9). One should select a point
Z_i' belonging to link C_i on the joint axis, which coincides with Z_i for
$q_i=0$. For $q_i \neq 0$, the distance between these points equals q_i. Vector $\vec{r}_{ii}$
is the vector between point Z_i at link C_{i-1} and the mass center of
link C_i. Therefore, it depends on q_i

$$\vec{r}_{ii} = \vec{r}_{ii}^{O} + q_i \vec{e}_i \qquad (1.3.1)$$

where $\vec{r}_{ii}^{O}$ is the distance between Z_i and O_i for $q_i=0$.

Beside the local coordinate frames attached to the links, a reference, fixed coordinate system Oxyz is to be assigned. With respect to this system external coordinates describing manipulation tasks, are defined. This system is usually positioned at manipulator base (Fig. 1.10). Its origin need not coincide with the center Z_1 of the first joint, so that a distance vector $\vec{r}_{o1}$ between these points has to be defined. This vector, or more exactly, its axis $\vec{e}_1$ serves as a reference for defining the zero value of joint coordinate q_1 in the same manner as vector $\vec{r}_{i-1,i}$ when defining joint coordinate q_i. If $\vec{r}_{o1}=0$ or $\vec{r}_{o1}$ is colinear with $\vec{e}_1$, an auxilliary vector has to be introduced, in the same way as it was described for vector $\vec{r}_{i-1,i}$.

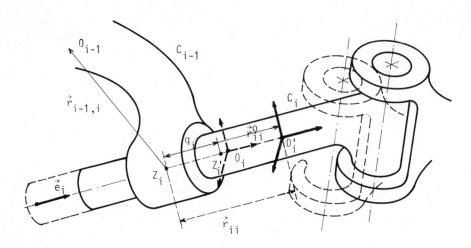

Fig. 1.9. Joint coordinate of a sliding kinematic pair

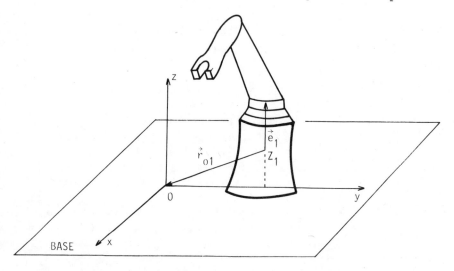

Fig. 1.10. Reference coordinate system

An illustration of the vectors defining manipulator links for a typical industrial robot presented in Fig. 1.11, is shown in Fig. 1.12. Here, the origin of the reference coordinate frame coincides with the center Z_1 of the first joint, resulting in $\vec{r}_{o1}=0$. Besides, $\vec{r}_{11}$ is parallel to $\vec{e}_1$. So, two auxilliary vectors $\vec{r}'_{o1}$ and $\vec{r}'_{11}$ are introduced in order to define position when $q_1=0$ (the case when the vectors become antiparallel). In Fig. 1.12 the arthropoid robot is shown in its initial configuration, when all joint coordinates are equal to zero. Kinematic parameters of all the links are listed in Table 1.1.

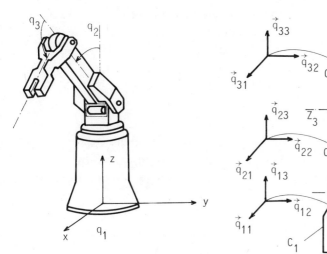

Fig. 1.11. Arthropoid robot

Fig. 1.12. Vectors defining the arthropoid robot in its initial position ($q_i=0$, $i=1,2,3$)

Link i	$\vec{r}_{ii}$	$\vec{r}_{i,i+1}$	$\vec{e}_i$	$\vec{e}_{i+1}$
1	(0, 0, 0.5)	(0, 0, -0.5)	(0, 0, 1)	(0, 1, 0)
2	(0, 0, 0.7)	(0, 0, -0.7)	(0, 1, 0)	(0, 1, 0)
3	(0, 0, 0.4)	(0, 0, -0.4)	(0, 1, 0)	-

$$\vec{r}'_{o1} = (-1, 0, 0), \qquad \vec{r}'_{11} = (0, 1, 0)$$

Table T.1.1. Kinematic parameters of the arthropoid robot

Transformation matrices

In order to describe the relationship between links, we have to determine the transformation matrices between link coordinate systems. These matrices transform vectors expressed with respect to one system into the vectors expressed in the second system. Having in mind that vectors describing the links are known with respect to their own link coordinate systems, we deduce that transformation matrices between these systems and the reference system have to be determined.

Let us first consider the computation of the transformation matrix from link coordinate system Q_i to system Q_{i-1}, assigned to two adjacent links. Denote this matrix by $^{i-1}A_i$. Matrix $^{i-1}A_i \in R^{3 \times 3}$ depends on kinematic parameters of the links and, for a revolute joint, on joint coordinate q_i. We will first consider the case when joint i is a revolute one.

In order to determine matrix $^{i-1}A_i$ for any value of joint coordinate q_i, we will first compute this matrix for $q_i = 0$, and then, by using the result obtained and Rodrigues formula we shall gain the desired matrix.

Consider links C_{i-1} and C_i when $q_i = 0$ (Fig. 1.13). Denote by $^{i-1}A_i^o$ the

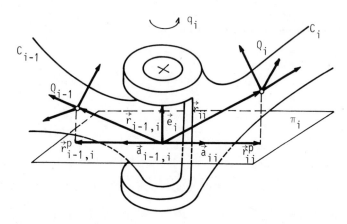

Fig. 1.13. Vectors relevant for obtaining transformation matrix $^{i-1}A_i$ for a revolute joint

matrix which transforms vectors expressed in the coordinate system of link i into the vectors expressed in system Q_{i-1}. For that we will use conditions of the type

$$\underset{\sim}{\vec{v}} = {}^{i-1}A_i^o \; \overset{\approx}{v}$$

where $\overset{\approx}{v}$ is a vector in system Q_i and $\underset{\sim}{\vec{v}}$ is the same vector expressed with respect to system Q_{i-1}. Evidently, for computing ${}^{i-1}A_i^o$ we need to know three linearly independant vectors expressed with respect to both the systems. Let us find three such vectors.

The vector which is known with certainty with respect to both the systems is the unit vector of the joint axis, i.e. vectors $\overset{\approx}{e}_i$ and $\underset{\sim}{\vec{e}}_i$ are known (defined as kinematic parameters of the kinematic chain). Besides, from the definition of the joint coordinate it follows that, for $q_i=0$, projections of vectors $\vec{r}_{ii}$ and $\vec{r}_{i-1,i}$ are antiparallel (Fig. 1.13). Denote by $\vec{a}_{ii}$ and $\vec{a}_{i-1,i}$ unit vectors corresponding to these projections. They are, evidently, obtained as

$$\vec{a}_{ii} = \frac{\vec{e}_i \times (\vec{r}_{ii} \times \vec{e}_i)}{|\vec{e}_i \times (\vec{r}_{ii} \times \vec{e}_i)|} \quad , \quad \vec{a}_{i-1,i} = \frac{\vec{e}_i \times (\vec{r}_{i-1,i} \times \vec{e}_i)}{|\vec{e}_i \times (\vec{r}_{ii} \times \vec{e}_i)|} \tag{1.3.2}$$

Since vector $\vec{r}_{ii}$ is known with respect to system Q_i, $\vec{a}_{ii}$ is also known with respect to Q_i, i.e. $\overset{\approx}{a}_{ii}$ is determined by (1.3.2). Vector $\vec{a}_{i-1,i}$ is, however, known with respect to Q_{i-1}, i.e. (1.3.2) gives as $\underset{\sim}{\vec{a}}_{i-1,i}$. Note that vectors $\vec{a}_{ii}$ and $\vec{a}_{i-1,i}$ are perpendicular to $\vec{e}_i$. Therefore, we can select $\vec{e}_i \times \vec{a}_i$ as the third vector needed for computing ${}^{i-1}A_i^o$. Thus, we have

$$\underset{\sim}{\vec{e}}_i = {}^{i-1}A_i^o \overset{\approx}{e}_i$$

$$\underset{\sim}{\vec{a}}_{i-1,i} = -{}^{i-1}A_i^c \overset{\approx}{a}_{ii} \tag{1.3.3}$$

$$\underset{\sim}{\vec{e}}_i \times \underset{\sim}{\vec{a}}_{i-1,i} = -{}^{i-1}A_i^o (\overset{\approx}{e}_i \times \overset{\approx}{a}_i)$$

Matrix ${}^{i-1}A_i^o$ is now completely defined as

$$^{i-1}A_i^o = [\underset{\sim}{\vec{e}}_i \;\; \underset{\sim}{\vec{a}}_{i-1,i} \;\; \underset{\sim}{\vec{e}}_i \times \underset{\sim}{\vec{a}}_{i-1,i}][\overset{\approx}{e}_i \;\; -\overset{\approx}{a}_{ii} \;\; -(\overset{\approx}{e}_i \times \overset{\approx}{a}_{ii})]^T \tag{1.3.4}$$

Note that matrix transpose instead of inverse has been used, due to the orthogonality of vectors $\vec{e}_i$, $\vec{a}_{ii}$ and $\vec{e}_i \times \vec{a}_{ii}$. Thus, the transformation matrix between two adjacent link coordinate frames, for $q_i=0$, has been determined.

If joint i is a sliding one, the transformation matrix between systems Q_i and Q_{i-1} does not depend on q_i, so that ${}^{i-1}A_i = {}^{i-1}A_i^o$ holds. Therefore, it is sufficient to determine matrix $^{i-1}A_i^o$ for a sliding kinematic pair. It can be obtained in a manner completely analogous to the above described method (Equations (1.3.3) - (1.3.4)). The only difference is that vectors $\vec{a}_{i-1,i}$ and $\underset{\sim}{\vec{a}}_{i-1,i}$ are to be explicitly defined as kinematic parameters (perpendicular to $\vec{e}_i$) (Fig. 1.14) but not to be computed from (1.3.2). While for a revolute joint these vectors determine the disposition of the links for $q_i=0$, for a sliding joint this disposition has to be explicitly imposed, since vectors $\vec{r}_{ii}$ and $\vec{r}_{i-1,i}$ do not define this disposition anyhow. Once the vectors $\underset{\sim}{\vec{a}}_{i-1,i}$ and $\vec{\tilde{a}}_{ii}$ are selected, the transformation matrix $^{i-1}A_i^o = {}^{i-1}A_i$ is obtained from Equations (1.3.3) and (1.3.4).

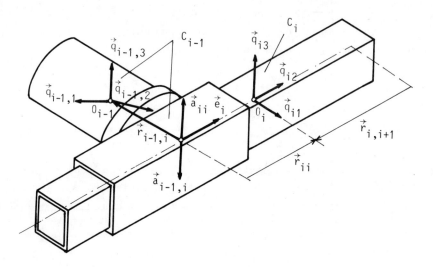

Fig. 1.14. Vectors relevant for obtaining transformation matrix $^{i-1}A_i$ for a sliding joint

Now we should determine the transformation matrix $^{i-1}A_i$ for an arbitrary q_i for a revolute joint. Let us first note that all vectors which describe link C_i are now known with respect to system Q_{i-1} if $q_i=0$, since matrix $^{i-1}A_i^o$ is determined. Thus, for example, we have $\vec{q}_{ij}^{o} =$ $= {}^{i-1}A_i^o\vec{q}_{ij}$, j=1,2,3. Since $\vec{q}_{i1} = [1\ 0\ 0]^T$, $\vec{q}_{i2} = [0\ 1\ 0]^T$ and $\vec{q}_{i3} = [0\ 0\ 1]^T$, it follows

$$^{i-1}A_i^o = [\vec{q}_{i1}^{o}\ \vec{q}_{i2}^{o}\ \vec{q}_{i3}^{o'}] \tag{1.3.5}$$

i.e. the columns of the transformation matrix correspond to the unit vectors of the coordinate system i, expressed with respect to system (i=1). The same relation evidently holds for $q_i \neq 0$ also, that is

$$^{i-1}A_i = [\vec{\underset{\sim}{q}}_{i1} \ \vec{\underset{\sim}{q}}_{i2} \ \vec{\underset{\sim}{q}}_{i3}] \tag{1.3.6}$$

It follows, therefrom, that, in order to obtain $^{i-1}A_i$, we have to determine projections of the unit vectors of the coordinate system Q_i to the system Q_{i-1}. Since vectors $\vec{\underset{\sim}{q}}_{ij}$ correspond to vector $\vec{\underset{\sim}{q}}^{\,o}_{ij}$ after a rotation by q_i, we will apply the finite rotations formula (Rodrigues formula) to obtain these vectors. Thus, we have

$$\vec{\underset{\sim}{q}}_{ij} = \vec{\underset{\sim}{q}}^{\,o}_{ij} \cos q_i + (1-\cos q_i)(\vec{\underset{\sim}{e}}_i \cdot \vec{\underset{\sim}{q}}^{\,o}_{ij})\vec{\underset{\sim}{e}}_i +$$
$$+ (\vec{\underset{\sim}{e}}_i \times \vec{\underset{\sim}{q}}^{\,o}_{ij}) \sin q_i, \qquad j=1,2,3 \tag{1.3.7}$$

The equations (1.3.6) - (1.3.7) completely determine the transformation matrix between two adjacent links interconnected by a revolute joint.

Once the matrix $^{i-1}A_i$ is obtained we can compute all the vectors describing the link i with respect to system Q_{i-1} for any q_i. For example

$$\vec{\underset{\sim}{r}}_{ii} = {}^{i-1}A_i \vec{r}_{ii}$$
$$\vec{\underset{\sim}{r}}_{i,i+1} = {}^{i-1}A_i \vec{r}_{i,i+1} \tag{1.3.8}$$

Equation (1.3.7) also indicates that each element of matrix $^{i-1}A_i$ can be presented as a linear combination of variables $\cos q_i$ and $\sin q_i$, with the coefficients being constant. Each element of $^{i-1}A_i = [a_{kj}]$ can be presented in the form

$$a_{kj} = a_{kj}^{(1)} \cos q_i + a_{kj}^{(2)} \sin q_i + a_{kj}^{(3)} \tag{1.3.9}$$

where coefficients $a_{kj}^{(\ell)}$, $\ell=1,2,3$ are constants for a given robot. These coefficients can be computed in advance. Very often, however, the structure of these matrix elements is considerably simpler (many of them are 0 or 1). This is due to the fact that for most industrial manipulators main axes of inertia (axes of the local coordinate systems) are parallel or perpendicular to joint axes.

Having obtained transformation matrices between adjacent link coordinate systems, $^{i-1}A_i$, $i=1,\ldots,n$ for all degrees of freedom, the transformation matrix from system Q_j to system Q_k, $k<j$, is simply obtain as

$$^k A_j = {}^k A_{k+1} \ {}^{k+1}A_{k+2} \ \cdots \ {}^{j-1}A_j, \qquad 0 \leqslant k < j \leqslant n \tag{1.3.10}$$

For k=0 we obtain the transformation matrix from coordinate system of link C_j to the fixed, reference coordinate system Oxyz.

If the transformation matrix $^O A_j$ between system Q_j and the reference system has been evaluated, the matrix $^O A_{j+1}$ betwen system Q_{j+1} and the reference system is obtained recursively as

$$^O A_{j+1} = {}^O A_j {}^j A_{j+1}, \qquad 1 < j < n-1 \tag{1.3.11}$$

Since matrix $^k A_j$ transforms vectors from system Q_j to system Q_k, its inverse $(^k A_j)^{-1}$ transforms vectors from system k to j and, therefore, represents matrix $^j A_k$. Since this matrix is also orthogonal, we can write

$$^j A_k = (^k A_j)^{-1} = {}^k A_j^T \tag{1.3.12}$$

Position vectors

Manipulator motion control requires information about manipulator link positions with respect to the reference coordinate system, as well as positions with respect to any link coordinate system.

We will, first, state a general problem of computing distance vectors $\vec{R}_{jk}$ betwen centers of any two joints k and j, k<j, expressed with respect to any link coordinate system Q_i, i=0,1,...,n. For i=0 we deal with the reference coordinate system (Fig. 1.15).

Vector $\vec{R}_{jk}$, from the center of joint k to the center of joint j, expressed with respect to system Q_i, i.e. vector $^i\vec{R}_{jk}$, can be obtained from the following equation

$$^i\vec{R}_{jk} = \sum_{\ell=k}^{j-1} {}^i\vec{R}_{\ell+1,\ell} \tag{1.3.13}$$

where $^i\vec{R}_{\ell+1,\ell}$ is the distance vector between center of joint ℓ to the center of joint $\ell+1$, expressed in the system of link i. This vector is specified by the geometry of the link ℓ and is given by

$$^i\vec{R}_{\ell+1,\ell} = {}^i A_\ell {}^\ell\vec{R}_{\ell+1,\ell} \tag{1.3.14}$$

where

$$\overset{\ell}{\vec{R}}_{\ell+1,\ell} = \overset{\sim}{\vec{R}}_{\ell+1,\ell} = \overset{\sim}{\vec{r}}_{\ell\ell} - \overset{\sim}{\vec{r}}_{\ell,\ell+1} \qquad (1.3.15)$$

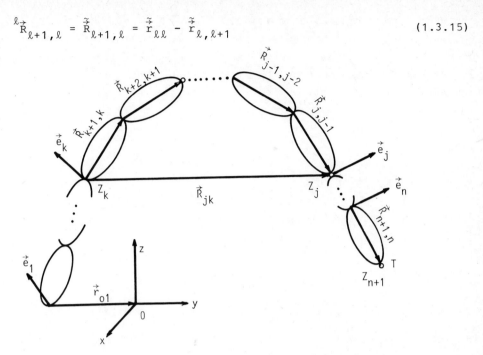

Fig. 1.15. Distance vectors between joint centers

For $\ell=n$, vector $\vec{R}_{n+1,n}$ represents the distance vector from the center of joint n to the manipulator tip (point T in Fig. 1.15).

Instead of Equation (1.3.13), one can employ the recursive relation

$$\overset{i}{\vec{R}}_{jk} = \overset{i}{\vec{R}}_{k+1,k} + \overset{i}{\vec{R}}_{j,k+1} \qquad (1.3.16)$$

or

$$\overset{i}{\vec{R}}_{jk} = \overset{i}{\vec{R}}_{j-1,k} + \overset{i}{\vec{R}}_{j,j-1} \qquad (1.3.17)$$

The following example will illustrate the use of these equations. Let us evaluate vectors between centers of all joints and manipulator tip, as well as the vector between the origin of the reference coordinate system and the manipulator tip. These vectors are to be expressed with respect to the reference system. We will assume that all transformation matrices between adjacent coordinate systems are known. According to Eq. (1.3.16) we have

$$\overset{i}{\vec{R}}_{n+1,i} = \overset{i}{\vec{R}}_{i+1,i} + \overset{i}{A}_{i+1} \overset{i+1}{\vec{R}}_{n+1,i+1}, \quad i=n-1,\dots,1 \qquad (1.3.18)$$

Applying this relation, successively for i=n-1, n-2,...,1 we obtain vectors between joint centers z_i and the manipulator tip, expressed with respect to coordinate systems Q_i. The same vectors, expressed with respect to the reference system are obtained from

$$^O\vec{R}_{n+1,i} = {}^O A_i \, {}^i\vec{R}_{n+1,i}, \qquad i=1,\ldots,n \qquad (1.3.19)$$

where

$$^O A_i = {}^O A_{i-1} \, {}^{i-1}A_i \qquad (1.3.20)$$

Manipulator hand position with respect to the reference, base coordinate system is now simply obtained as

$$\vec{r}_p = -\vec{r}_{o,1} + {}^O\vec{R}_{n+1,1} = [x \; y \; z]^T \qquad (1.3.21)$$

where $\vec{r}_{o1}$ is the vector between the center of joint 1 and the origin of the reference system.

Similarly, the same vectors can be obtained with respect to coordinate system Q_n of the final link, i.e. the gripper. As in the previous case, we evaluate $^i\vec{R}_{n+1,i}$ using (1.3.18), and then apply

$$^n\vec{R}_{n+1,i} = {}^n A_i \, {}^i\vec{R}_{n+1,i} = {}^i A_n^T \, {}^i\vec{R}_{n+1,i}$$

$$^i A_n = {}^i A_{i+1} \, {}^{i+1}A_n, \qquad i = n-1,\ldots,1 \qquad (1.3.22)$$

Thus we have obtained all the position vectors needed for kinematic modelling of manipulators.

1.3.2. Homogeneous transformations

Another method for kinematic modelling of manipulators, which is widely accepted today, is based on Denavit-Hartenberg notation [1, 8, 9]. Unlike the Rodrigues formula approach, it was not developed to match dynamic modelling demands, but aimed solely for the kinematic analysis of mechanism. A correlation between these two kinematic modelling techniques, with respect to generality, complexity and convenience for dynamic modelling have been discussed in [10] in detail.

Let us now consider how link coordinate systems are assigned to links by this method, together with kinematic link parameters which are here introduced.

Consider a simple open kinematic chain with n links. Each link is characterized by two dimensions: the common normal distance a_i (along the common normal between axes of joint i and i+1), and the twist angle α_i between these axes in the plane perpendicular to a_i.

Each joint axis has two normals to it a_{i-1} and a_i (Fig. 1.16). The relative position of these normals along the axis of joint i is given by d_i.

Denote by $0_i x_i y_i z_i$ the local coordinate system assigned to link i. We will first consider revolute joints (Fig. 1.6). The origin of the coordinate frame of link i is set to be at the intersection of the common normal between the axis of joint i+1. In the case of intersecting joint axes, the origin is set to be at the point of intersection of the joint axes. If the axes are parallel, the origin is chosen to make the joint distance zero for the next link whose coordinate origin is defined. The axes of the link coordinate system $0_i x_i y_i z_i$ are to be selected in the following way. The $\vec{z}_i$ axis of system i should coincide

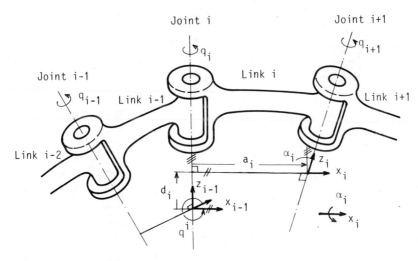

Fig. 1.16. Denavit-Hartenberg kinematic parameters for a revolute kinematic pair

with the axis of joint i+1, about which rotation q_{i+1} is performed. The $\vec{x}_i$ axis will be aligned with any common normal which exists (usually the normal betwen axes of joint i and i+1) and is directed from joint i to joint i+1. In the case of intersecting joint axes, the $\vec{x}_i$ axis is chosen to be parallel or antiparallel to the vector cross product $\vec{z}_{i-1} \times \vec{z}_i$. The $\vec{y}_i$ axis satisfies $\vec{x}_i \times \vec{y}_i = \vec{z}_i$.

The joint coordinate q_i for a revolute joint is now defined as the angle between axes $\vec{x}_{i-1}$ and $\vec{x}_i$ (Fig. 1.16). It is zero when these axes are parallel and have the same direction. The twist angle α_i is measured from axis $\vec{z}_{i-1}$ to $\vec{z}_i$, i.e. as a rotation about $\vec{x}_i$ axis.

Let us now consider prismatic joints (Fig. 1.17). Here, the distance d_i becomes joint variable q_i, while parameter a_i has no meaning and is set to zero. The origin of coordinate system corresponding to the sliding joint i is chosen to coincide with the next defined link origin. The $\vec{z}_i$ axis is aligned with the axis of joint i+1. The $\vec{x}_i$ axis is chosen to be parallel or antiparallel to vector $\vec{z}_{i-1} \times \vec{z}_i$. Zero joint coordinate $q_i = d_i$ is defined when origins of systems $O_{i-1} x_{i-1} y_{i-1} z_{i-1}$ and $O_i x_i y_i z_i$ coincide. For a prismatic joint, the angle θ_i between axis $\vec{x}_{i-1}$ and $\vec{x}_i$ is fixed and represents a kinematic parameter together with the twist angle α_i.

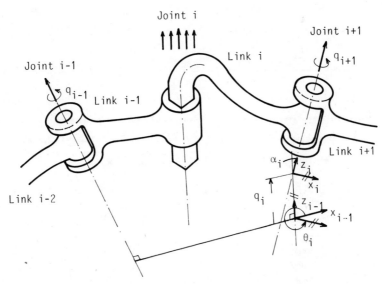

Fig. 1.17. Denavit-Hartenberg kinematic parameters for a prismatic joint

The origin of the reference coordinate system Oxyz is set to be coincident with the origin of the first link 1. Zero value of joint coordinate q_1 occurs when axes $\vec{x}$ and $\vec{x}_1$ are parallel and have the same directions.

Homogeneous transformation matrices

Let us now determine transformation matrices between link coordinate systems and distance vectors between their origins. The homogeneous transformation matrices involve both the information in the form of 4×4 matrices.

Denote by $^{i-1}T_i$ the homogeneous transformation relating between the coordinate system of link i and the coordinate frame of link i-1. It can be presented in the form

$$^{i-1}T_i = \left[\begin{array}{c|c} {}^{i-1}A_i & {}^{i-1}\vec{R}_{i,i-1} \\ \hline 0\ 0\ 0 & 1 \end{array} \right] \in R^{4 \times 4} \tag{1.3.23}$$

where $^{i-1}A_i \in R^{3 \times 3}$ is the matirx which transforms vectors form system i into system i-1, and $^{i-1}\vec{R}_{i,i-1}$ is the distance vector between origins of systems i-1 and i, expressed with respect to system i-1. Since matrix $^{i-1}A_i$ describes rotation between these systems it is sometimes referred to as rotation matrix or orientation matrix.

The transformation of system i-1 into system i can be described by the following set of rotations and translations: rotation about $\vec{z}_{i-1}$ by angle q_i (for a sliding joint by θ_i), translation along $\vec{z}_{i-1}$ for d_i (q_i - for a sliding joint) translation along rotated $\vec{x}_{i-1} = \vec{x}_i$ for a_i and rotation about $\vec{x}_i$ for the twist angle α_i. This sequence of rotations and translations can be presented as a product of the following homogeneous transformation matrices

$$^{i-1}T_i = \begin{bmatrix} \cos q_i & -\sin q_i & 0 & 0 \\ \sin q_i & \cos q_i & 0 & 0 \\ 0 & 0 & 1 & 0 \\ 0 & 0 & 0 & 1 \end{bmatrix} \begin{bmatrix} 1 & 0 & 0 & a_i \\ 0 & 1 & 0 & 0 \\ 0 & 0 & 1 & d_i \\ 0 & 0 & 0 & 1 \end{bmatrix}$$

$$\begin{bmatrix} 1 & 0 & 0 & 0 \\ 0 & \cos \alpha_i & -\sin \alpha_i & 0 \\ 0 & \sin \alpha_i & \cos \alpha_i & 0 \\ 0 & 0 & 0 & 1 \end{bmatrix} \tag{1.3.24}$$

This relation yields the homogeneous transformation matrix between two successive coordinate systems i and i-1 for a revolute joint

$$
{}^{i-1}T_i = \begin{bmatrix} \cos q_i & -\sin q_i \cos \alpha_i & \sin q_i \sin \alpha_i & a_i \cos q_i \\ \sin q_i & \cos q_i \cos \alpha_i & -\cos q_i \sin \alpha_i & a_i \sin q_i \\ 0 & \sin \alpha_i & \cos \alpha_i & d_i \\ 0 & 0 & 0 & 1 \end{bmatrix}
$$

$$(1.3.25)$$

For a prismatic joint it becomes

$$
{}^{i-1}T_i = \begin{bmatrix} \cos \theta_i & -\sin \theta_i \cos \alpha_i & \sin \theta_i \sin \alpha_i & 0 \\ \sin \theta_i & \cos \theta_i \cos \alpha_i & -\cos \theta_i \sin \alpha_i & 0 \\ 0 & \sin \alpha_i & \cos \alpha_i & q_i \\ 0 & 0 & 0 & 1 \end{bmatrix} \quad (1.3.26)
$$

Having obtained the homogeneous transformation matrices between sucessive coordinate frames, ${}^{i-1}T_i$, i=1,...,n, one can easily obtain the homogeneous transformation between any two systems j and k, k<j from

$$
{}^kT_j = {}^kT_{k+1} \; {}^{k+1}T_{k+2} \; \cdots \; {}^{j-1}T_j, \qquad 0 \leqslant k < j \leqslant n \tag{1.3.27}
$$

It gives information on the rotation between systems j and k and the distance vector between their origins. For k=0 and j=n this matrix gives the Cartesian coordinates of the manipulator tip and information on hand orientation

$$
{}^0T_n = \left[\begin{array}{c|c} {}^0A_n & \begin{matrix} x \\ y \\ z \end{matrix} \\ \hline 0 \; 0 \; 0 & 1 \end{array} \right]
$$

$$(1.3.28)$$

with respect to the base reference coordinate frame.

If the homogeneous transformation matrix T between two coordinate frames is known

$$
T = \left[\begin{array}{c|c} A & p \\ \hline 0 \; 0 \; 0 & 1 \end{array} \right]
$$

where $A \in R^{3 \times 3}$, $p \in R^3$, then its inverse transformation is simply obtained as

$$
T^{-1} = \left[\begin{array}{ccc|c} & & & -p \cdot a^{(1)} \\ & A^T & & -p \cdot a^{(2)} \\ & & & -p \cdot a^{(3)} \\ \hline 0 & 0 & 0 & 1 \end{array} \right]
$$

where $\cdot$ denotes vector scalar product, $a^{(1)}$, $a^{(2)}$ and $a^{(3)}$ are the 1., 2. and 3. columns of matrix A. Here the orthogonality property of matrix A is used.

1.3.3. Spherical coordinates

Let us now consider computation of manipulator hand position specified by spherical coordinates r, α and β (Eq. 1.2.3, Fig. 1.3). We could see from the previous two subsections that kinematic modelling techniques yield Cartesian coordinates x, y, z of the manipulator hand. We will use these coordinates to obtain spherical coordinates. Given a vector of joint coordinates $q = [q_1 \cdots q_n]^T \epsilon Q$, Cartesian coordinates are computed first, followed by spherical coordinates.

The relationship between Cartesian coordinates x, y and z and translation r and rotation angles α and β, is evident from Fig. 1.3

$$x = r \cos \alpha \sin \beta \qquad\qquad (1.3.29)$$

$$y = r \sin \alpha \sin \beta \qquad\qquad (1.3.30)$$

$$z = r \cos \beta \qquad\qquad (1.3.31)$$

In order to obtain α we will multiply both sides of Eq. (1.3.29) by $\sin \alpha$ and both sides of (1.3.30) by $\cos \alpha$, and then subtract these two equations. Thus we obtain

$$x \sin \alpha - y \cos \alpha = 0 \qquad\qquad (1.3.32)$$

and

$$\alpha = \text{arctg} \frac{y}{x} + k \Pi \qquad\qquad (1.3.33)$$

Coordinate β can be obtained in the following way: by multiplying Eq.

(1.3.29) by cos α and Eq. (1.3.30) by sin α, and adding these equations, we obtain

$$x \cos \alpha + y \sin \alpha = r \sin \beta \qquad (1.3.34)$$

The above relation together with (1.3.31) defines angle β

$$\beta = \text{arctg}((x \cos \alpha + y \sin \alpha)/z) + 2k\Pi \qquad (1.3.35)$$

The singular case when r=0, never occurs for practical manipulators, since the reference coordinate system is usually centered at the center of first joint.

Multiplying both sides of Eq. (1.3.34) by sin β and both sides of (1.3.31) by cos β and adding these two equations, we obtain

$$r = (x \cos \alpha + y \sin \alpha) \sin \beta + z \cos \beta \qquad (1.3.35)$$

In order to avoid problems occuring when denominators in Equations (1.3.33) and (1.3.35) are small, one should use the computer available function ATAN2(x, y) which returns angles in the range $\alpha \in [-\Pi, \Pi)$, depending on signs of x and y. Only in the case x=y=0, the solution is not uniquely defined.

However, the indefinitness of the 2kΠ or kΠ type cannot be avoided by the use of this function. Namely, manipulation tasks are described as desired continuous trajectories within the external coordinate space. Therefore, the real values of external coordinates, computed for a given vector of the joint coordinates q, must not be changed abruptly. For example, computed values α and β should never change for 2Π rad in two successive time intervals during motion execution. This means that the direct kinematic problem solution must not yield great changes in external coordinates (angles) for small changes in joint coordinates. This would certainly happen if angles (α, β or ψ, θ, φ) were calculated in the range [-Π, Π), and in cases in which the manipulator hand makes a full circle during it's motion. The abrupt changes of the computed external coordinates and consequently the velocities, would cause difficult numerical problems in motion synthesis, especially if elimination of linearization error is desired (see Section 3.3, Equation (3.3.7)).

Therefore, it is desirable to select one from the set of feasible solutions, according to a given criterion. For example, the solution nearest to the desired value of the same angle imposed by the task may be chosen. Alternatively, one may choose the solution nearest to the previous value of the same coordinate. The index k set is not that large as it may appear, since for real industrial manipulators, the motion of the joints is mechanically constrained. Thus, index k usually takes values -1, 0, or 1.

In the case when one of the angles is physically nonuniquely defined (e.g. α is not defined if x=y=0), we can also adopt the angle so that it takes the value from the desired trajectory or from the previous time instant. This discussion is not related only to spherical coordinates α and β, but can also be applied to cylindrical coordinate α and Euler angles specifying manipulator hand orientation.

1.3.4. Cylindrical coordinates

Calculation of cylindrical coordinates, introduced in Sect. 1.2 by Eq. (1.2.4), Fig. 1.3, will also be carried out using Cartesian coordinates x, y and z. The relationship between them is given by

$$x = r \cos \alpha \tag{1.3.36}$$

$$y = r \sin \alpha \tag{1.3.37}$$

$$z = z \tag{1.3.38}$$

The cylindrical z coordinate coincides with the Cartesian z coordinate, while α is obtained multiplying both sides of (1.3.36) by $\sin \alpha$ and (1.3.37) by $\cos \alpha$ and subtracting these equations

$$x \sin \alpha - y \cos \alpha = 0 \tag{1.3.39}$$

this gives

$$\alpha = \text{arctg} \, \frac{y}{x} + k\Pi \tag{1.3.40}$$

Having computed α, we can obtain r by multiplying Eq. (1.3.36) by $\cos \alpha$ and Eq. (1.3.37) by $\sin \alpha$ and adding these equations

$$r = x \cos \alpha + y \sin \alpha \tag{1.3.41}$$

Thus, all the types of external coordinates describing the manipulator
hand position have been determined.

1.4. Manipulator Hand Velocities

In order to make a complete specification of manipulation tasks a description of manipulator hand orientation with respect to objects in
the manipulator work space is necessary.

Having assigned a coordinate frame to the last link (link n) and a reference coordinate system, this orientation is completely defined by the
transformation matrix ^OA_n between these two systems. As we saw in Subsections 1.3.1 and 1.3.2 this matrix is obtained by kinematic modelling
procedures. However, this orientation can be described by means of three
variables only instead of the complete matrix $^OA_n \in R^{3 \times 3}$. These variables
are either a type of Euler angles or Euler parameters. These variables
form the second part of the external coordinates vector introduced in
Sect. 1.2, Eqations (1.2.5), (1.2.6). Computation of these variables,
given a vector of joint coordinates q, represents a part of the direct
problem solution.

In motion synthesis, however, the manipulator position and orientation
is frequently stored in the form of complete homogeneous transformations ^OT_n. This is convenient in the case when the inverse kinematic
problem can be solved analytically, so that all the matrix elements are
required for the solution. Besides, if the manipulation task is specified with respect to a moving coordinate frame, and on-line multiplication of homogeneous transformations is required, it is also convenient
to have these matrices already memorized.

In this book we will consider the specification of orientation by three
variables. We will first deal with Euler angles ψ, θ and φ which correspond to yaw, pitch and roll. We will then consider Euler parameters,
although they have not been used extensively so far.

1.4.1. Euler angles

Let us consider two coordinate frames and describe rotation between
them. The first system Q_n is assigned to the last manipulator link, and
its unit axes are denoted by $\vec{q}_{n1}$, $\vec{q}_{n2}$ and $\vec{q}_{n3}$ (Fig. 1.18). The second

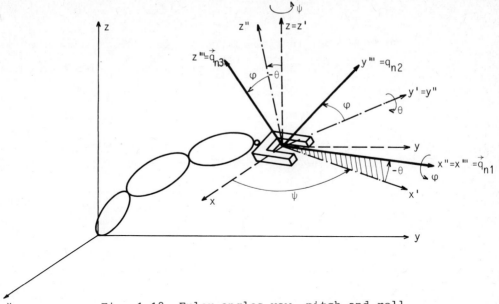

Fig. 1.18. Euler angles yaw, pitch and roll

system is the reference system Oxyz. The transformation matrix $^O A_n$ which maps vectors from system Q_n to system Oxyz is

$$
^O A_n = [\, ^O \vec{q}_{n1} \quad ^O \vec{q}_{n2} \quad ^O \vec{q}_{n3} \,] = \begin{bmatrix} a_{11} & a_{12} & a_{13} \\ a_{21} & a_{22} & a_{23} \\ a_{31} & a_{32} & a_{33} \end{bmatrix} \tag{1.4.1}
$$

We will assume that, given a joint coordinate vector, the numerical values a_{ij}, $i,j = 1,2,3$ hawe been computed. The problem is to determine the Euler angles - yaw, pitch and roll, which correspond to a certain matrix $^O A_n = [a_{ij}]$.

The rotation of system Q_n with respect to Oxyz can be described by the following sequence of rotations: a rotation by ψ about z axis (yaw), followed by a rotation by θ about the new y axis (pitch), and rotation by φ about the new x axis (roll) (Fig. 1.18). This sequence of rotations corresponds to the following product of transformation matrices

$$
^O A_n = \begin{bmatrix} \cos\psi & -\sin\psi & 0 \\ \sin\psi & \cos\psi & 0 \\ 0 & 0 & 1 \end{bmatrix} \begin{bmatrix} \cos\theta & 0 & \sin\theta \\ 0 & 1 & 0 \\ -\sin\theta & 0 & \cos\theta \end{bmatrix} \begin{bmatrix} 1 & 0 & 0 \\ 0 & \cos\varphi & -\sin\varphi \\ 0 & \sin\varphi & \cos\varphi \end{bmatrix}
$$
$$
\tag{1.4.2}
$$

or

$$
{}^{O}A_{n} = \begin{bmatrix} \cos\psi\cos\theta & \cos\psi\sin\theta\sin\varphi-\sin\psi\cos\varphi & \cos\psi\sin\theta\cos\varphi+\sin\psi\sin\varphi \\ \sin\psi\cos\theta & \sin\psi\sin\theta\sin\varphi+\cos\psi\cos\varphi & \sin\psi\sin\theta\cos\varphi-\cos\psi\sin\varphi \\ -\sin\theta & \cos\theta\sin\varphi & \cos\theta\cos\varphi \end{bmatrix}
$$

$$(1.4.3)$$

Equating the numerical values of a_{ij} elements (1.4.1), obtained by a kinematic modelling procedure, with the elements from (1.4.3), we have

$$a_{11} = \cos \psi \cos \theta \qquad (1.4.4)$$

$$a_{21} = \sin \psi \cos \theta \qquad (1.4.5)$$

$$a_{31} = -\sin \theta \qquad (1.4.6)$$

$$a_{12} = \cos \psi \sin \theta \sin \varphi -\sin \psi \cos \varphi \qquad (1.4.7)$$

$$a_{22} = \sin \psi \sin \theta \sin \varphi +\cos \psi \cos \varphi \qquad (1.4.8)$$

$$a_{32} = \cos \theta \sin \varphi \qquad (1.4.9)$$

$$a_{13} = \cos \psi \sin \theta \cos \varphi +\sin \psi \sin \varphi \qquad (1.4.10)$$

$$a_{23} = \sin \psi \sin \theta \cos \varphi -\cos \psi \sin \varphi \qquad (1.4.11)$$

$$a_{33} = \cos \theta \cos \varphi \qquad (1.4.12)$$

Yaw angle ψ can be obtained in the following way. Multiplying both sides of Equation (1.4.4) by $\sin \psi$ and both sides of (1.4.5) by $\cos \psi$, and subtracting these two equations, we obtain

$$a_{11}\sin \psi -a_{21}\cos \psi = 0 \qquad (1.4.13)$$

or

$$\psi = \text{arctg} \frac{a_{21}}{a_{11}} + k\Pi \qquad (1.4.14)$$

Pitch angle θ can be computed by multiplying (1.4.4) by $\cos \psi$ and (1.4.5) by $\sin \psi$ and adding these two equations

$$a_{11}\cos \psi + a_{21}\sin \psi = \cos \theta \qquad (1.4.15)$$

The above equation together with (1.4.6) yields

$$\theta = \text{arctg}[-a_{31}/(a_{11}\cos\psi + a_{21}\sin\psi)] + 2k\Pi \qquad (1.4.16)$$

The roll angle is obtained in a similar way. Multiplying (1.4.10) by $\sin\psi$ and (1.4.11) by $\cos\psi$, and subtracting these two equations we get

$$a_{13}\sin\psi - a_{23}\cos\psi = \sin\varphi \qquad (1.4.17)$$

Similarly, by multiplying (1.4.7) by $-\sin\psi$ and (1.4.8) by $\cos\psi$ and adding these two equations we have

$$-a_{12}\sin\psi + a_{22}\cos\psi = \cos\varphi \qquad (1.4.18)$$

The above equations give the solution to the roll angle

$$\varphi = \text{arctg}\frac{a_{13}\sin\psi - a_{23}\cos\psi}{-a_{12}\sin\psi + a_{22}\cos\psi} + 2k\Pi \qquad (1.4.19)$$

As discussed in Subsection 1.3.3 (see the comments following Equation (1.3.35)), problems occuring when denominators in Equations (1.4.14), (1.4.16) and (1.4.19) are small, may be avoided by the use of computer function ATAN2. The selection of the value of index k is performed in the same way as described in Subsection 1.3.3. The only singular situation is when $\theta = k\Pi/2$, since in this case $\cos\theta = 0$, resulting in $a_{11} = a_{21} = 0$. This occurs when the manipulator hand is pointing straight up or down and both yaw and roll correspond to the same rotation. Equation (1.4.14) yields no solution to ψ. In this case the value for ψ can be chosen arbitrarily. We suggest that the value for ψ should be the same as the desired value imposed by external coordinate trajectory, in that time instant. Once the ψ angle is chosen the φ is computed from (1.4.19).

1.4.2. Euler parameters

The specification of a rigid body orientation by means of Euler parameters was introduced in papers [11-13], and applied to the specification of manipulator hand orientation [4].

If the rotation of the coordinate system Q_n of the last link respect to the reference system is specified by a rotation angle ε about a unit axis of rotation $\vec{e}$, then Euler parameters are

$$\lambda_1 = e_1 \sin \frac{\varepsilon}{2}$$

$$\lambda_2 = e_2 \sin \frac{\varepsilon}{2} \qquad (1.4.20)$$

$$\lambda_3 = e_3 \sin \frac{\varepsilon}{2}$$

and $\lambda_o = \cos \frac{\varepsilon}{2}$, where e_i, $i=1,2,3$ represent projections of vector $\vec{e}$ to the reference coordinate system. It is always possible to choose $\varepsilon \in [0, \Pi]$. It can be shown that relations

$$\varepsilon = 2 \text{Arccos} \sqrt{1-(\lambda_1^2 + \lambda_2^2 + \lambda_3^2)}$$

$$e_i = \frac{\lambda_i}{\sqrt{\lambda_1^2 + \lambda_2^2 + \lambda_3^2}} \quad , \qquad i=1,2,3, \qquad \lambda_{1,2,3} \neq 0$$

hold. When $\lambda_i = 0$, $i=1,2,3$ we choose e_i according to convention

$$e_i = \frac{1}{\sqrt{3}} \quad , \qquad i=1,2,3$$

The transformation matrix ^OA_n between these two systems can now be written in the form

$$^OA_n = \begin{bmatrix} 2(\lambda_o^2 + \lambda_1^2) - 1 & 2(\lambda_1\lambda_2 - \lambda_o\lambda_3) & 2(\lambda_1\lambda_3 + \lambda_o\lambda_2) \\ 2(\lambda_1\lambda_2 + \lambda_o\lambda_3) & 2(\lambda_o^2 + \lambda_2^2) - 1 & 2(\lambda_2\lambda_3 - \lambda_o\lambda_1) \\ 2(\lambda_1\lambda_3 - \lambda_o\lambda_2) & 2(\lambda_2\lambda_3 + \lambda_o\lambda_1) & 2(\lambda_o^2 + \lambda_3^2) - 1 \end{bmatrix} \qquad (1.4.21)$$

By equating the elements of the matrix on the right-hand side of the above equation with the elements of ^OA_n obtained in the kinematic modelling procedure, we can obtain Euler parameters as

$$\lambda_1 = \frac{1}{2} \text{sgn}(a_{32} - a_{23}) \sqrt{(a_{11} - a_{22} - a_{33} + 1)}$$

$$\lambda_2 = \frac{1}{2} \text{sgn}(a_{13} - a_{31}) \sqrt{(-a_{11} + a_{22} - a_{33} + 1)} \qquad (1.4.22)$$

$$\lambda_3 = \frac{1}{2} \text{sgn}(a_{21} - a_{12}) \sqrt{(-a_{11} - a_{22} + a_{33} + 1)}$$

where $^OA_n = [a_{ij}]$.

Having obtained the equations for computing the manipulator hand position specified by Cartesian, spherical or cylindrical coordinates, and hand orientation by means of Euler angles and parameters, given a vector of joint coordinates, we have solved the direct kinematic problem $x_e = f(q)$.

1.5. Manipulator Hand Velocities

Another important problem, which has to be solved in manipulator motion synthesis, is establishing the relationship between the linear and angular velocities of the manipulator hand and the joint coordinate velocities. This relationship is essential for manipualtor rate control, as well as for iterative procedures for inverse kinematic problem solving, i.e. computing joint coordinates which correspond to a given manipulator hand position and orientation.

We will first derive relations for computing the linear and angular velocities of manipulator link i, assuming that the velocities of link (i-1) are already known. Using these recursive relations we will derive nonrecursive relations, which give an explicit relationship between link velocities and joint rates. Thus, the Cartesian coordinate velocities of the manipulator hand will be related to joint rates, and the angular velocities of hand rotation about the axes of the reference coordinate frame will be related to joint rates.

Since manipulator hand position and orientation can be specified by various types of external coordinates, it is convenient to establish the direct relationship between these external coordinate velocities and joint rates. We will derive expressions for Jacobian matrices (or submatrices) which correspond to Cartesian coordinates, spherical coordinates, cylindrical coordinates, Euler angles and Euler parameters.

It is sometimes useful to move the manipulator hand in the directions of it's axes and to rotate it about these axes. Therefore, we will also derive the Jacobian which corresponds to linear and angular velocities with respect to the coordinate system of the last link.

1.5.1. <u>Recursive and nonrecursive relations</u>
<u>for linear and angular velocities</u>

Let us consider a serial-link manipulator consisting of n links and n joints. We will assume that a vector of joint coordinates q - is given and that the first phase of kinematic modelling was completed, yielding position vectors of the links with respect to the reference frame, as well as transformation matrices between the links. We will aslo assume that the vector of joint rates $\dot{q} = [\dot{q}_1 \cdots \dot{q}_n]^T$ is given.

Angular velocity

We will first determine the vector of angular velocity $\vec{\omega}_i$ of the link i with respect to the reference frame (Fig. 1.19). This vector can be expressed as the sum of the angular velocity $\vec{\omega}_{i-1}$ of the previous link and the vector or relative angular velocity between links i-1 and i. For a revolute joint we can write

$$\vec{\omega}_i = \vec{\omega}_{i-1} + \dot{q}_i \vec{e}_i, \qquad 1 \leqslant i \leqslant n \tag{1.5.1}$$

where $\vec{e}_i$ is the unit vector of joint axis i. All the points belonging to link C_i have the same angular velocity. Equation (1.5.1) gives the angular velocity vector for link i, expressed with respect to any co-ordinate system, if the joint axes $\vec{e}_i$, i=1,...,n are expressed with respect to that system.

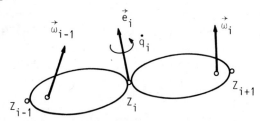

Fig. 1.19. Angular velocity of link i

In the case of a sliding joint, relative rotation between the links does not exist, so that we have

$$\vec{\omega}_i = \vec{\omega}_{i-1}, \qquad 1 \leqslant i \leqslant n \tag{1.5.2}$$

Equations (1.5.1) and (1.5.2) can be presented in the form of a single relation

$$\vec{\omega}_i = \vec{\omega}_{i-1} + \dot{q}_i \vec{e}_i (1-\xi_i), \qquad i=1,\ldots,n \qquad (1.5.3)$$

where ξ_i indicates whether the joint i is a revolute one ($\xi_i=0$), or a sliding one ($\xi_i=1$). We will assume that the angular velocity of the manipulator base equals zero, i.e. $\vec{\omega}_o=0$.

An equivalent, but nonrecursive relation for the angular velocity of link i is obtained by summing expressions (1.5.3)

$$\vec{\omega}_i = \sum_{j=1}^{i} \dot{q}_j \vec{e}_j (1-\xi_j) \qquad (1.5.4)$$

This equation can be presented in the following matrix form

$$\vec{\omega}_i = [\vec{e}_1(1-\xi_1) \mid \cdots \mid \vec{e}_i(1-\xi_i \mid 0 \cdots 0]\dot{q} \qquad (1.5.5)$$

where $\dot{q} = [\dot{q}_1 \dot{q}_2 \cdots \dot{q}_n]^T \epsilon R^n$, and the matrix on the right-hand side of (1.5.5) is an $3 \times n$ matrix.

For i=n Equation (1.5.5) gives angular velocity of the last link. Since the last link destines the orientation of the work object, and its angular velocity is the angular velocity of the work object, vector $\vec{\omega}_n$ has special importance

$$\vec{\omega}_n = J_\omega \dot{q} \qquad (1.5.6)$$

where, according to (1.5.5), J_ω is given by

$$J_\omega = [\vec{e}_1(1-\xi_1) \mid \cdots \mid \vec{e}_n(1-\xi_n)] \epsilon R^{3 \times n} \qquad (1.5.7)$$

If vectors $\vec{e}_i$, i=1,...,n are expressed with respect to the base reference frame, Eq. (1.5.6) gives the angular velocity of the last link with respect to the same frame. If $\vec{e}_i$ are expressed with respect to the coordinate system of the last link, Eq. (1.5.6) gives the angular velocity of the last link with respect to the end-effector coordinate frame.

Linear velocity

Let us now consider the translational velocity of manipulator links. Since the links have rotational motion beside the linear motion, the linear velocity of different points belonging to the same link is not

the same. It is well known from the rigid body kinematics that the
linear velocity of a point B is equal to the sum of the linear velocity of a point A and its relative rotation about this point

$$\vec{v}_B = \vec{v}_A + \vec{\omega} \times \vec{r}_{AB}$$

We will use this theorem to determine the linear velocities of the
end points of the links, i.e. the centers of joints Z_i, $i=1,\ldots,n+1$.
Thus we can write the following recursive relation for a revolute joint
directly

$$\vec{v}_i = \vec{v}_{i-1} + \vec{\omega}_i \times \vec{R}_{i+1,i} \qquad\qquad (1.5.8)$$

where $\vec{v}_i$ is the linear velocity of the end point of link i (point Z_{i+1}),
$\vec{v}_{i-1}$ - is the linear velocity of the end point of link (i-1) (point
Z_i), $\vec{\omega}_i$ - the angular velocity of link i, $\vec{R}_{i+1,i}$ - the distance vector
between points Z_i and Z_{i+1} (Fig. 1.20).

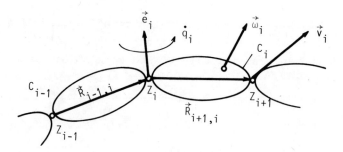

Fig. 1.20. Linear velocity $\vec{v}_i$

In the case of a prismatic joint, an additional term should be included,
which originates from the change of the joint coordinate itself

$$\vec{v}_i = \vec{v}_{i-1} + \vec{\omega}_i \times \vec{R}_{i+1,i} + \dot{q}_i \vec{e}_i \qquad\qquad (1.5.9)$$

Equations (1.5.8) and (1.5.9) can be presented in a summarized form

$$\vec{v}_i = \vec{v}_{i-1} + \vec{\omega}_i \times \vec{R}_{i+1,i} + \dot{q}_i \vec{e}_i \xi_i, \qquad i=1,\ldots,n \qquad (1.5.10)$$

where $\xi_i=0$ for revolute joints and $\xi_i=1$ for prismatic joints. Vector
$\vec{v}_o$ corresponds to the linear velocity of the manipulator base. We will
assume $\vec{v}_o=0$.

Let us now derive nonrecursive expressions for the linear velocity of the end point of link i. According to (1.5.10) we have

$$\vec{v}_i = \sum_{j=1}^{i} (\vec{\omega}_j \times \vec{R}_{j+1,j} + \dot{q}_j \vec{e}_j \xi_j) \tag{1.5.11}$$

Replacing $\vec{\omega}_j$ by (1.5.4) we obtain

$$\vec{v}_i = \sum_{j=1}^{i} [\sum_{k=1}^{j} (\vec{e}_k \times \vec{R}_{j+1,j}) (1-\xi_k) \dot{q}_k + \dot{q}_j \vec{e}_j \xi_j] \tag{1.5.12}$$

or

$$\vec{v}_i = \sum_{j=1}^{i} [\sum_{k=1}^{j} \vec{c}_{jk} \dot{q}_k + \dot{q}_j \vec{e}_j \xi_j] \tag{1.5.13}$$

where

$$\vec{c}_{jk} = (\vec{e}_k \times \vec{R}_{j+1,j}) (1-\xi_k)$$

By rearranging the order of summing in (1.5.13) we get

$$\vec{v}_i = \sum_{j=1}^{i} (\sum_{k=j}^{i} \vec{c}_{kj} + \vec{e}_j \xi_j) \dot{q}_j =$$

$$= \sum_{j=1}^{i} [\sum_{k=j}^{i} (\vec{e}_j \times \vec{R}_{k+1,k}) (1-\xi_j) + \vec{e}_j \xi_j] \dot{q}_j =$$

$$= \sum_{j=1}^{i} [\vec{e}_j (1-\xi_j) \times (\sum_{k=j}^{i} \vec{R}_{k+1,k}) + \vec{e}_j \xi_j] \dot{q}_j =$$

$$= \sum_{j=1}^{i} [(\vec{e}_j \times \vec{R}_{i+1,j}) (1-\xi_j) + \vec{e}_j \xi_j] \dot{q}_j \tag{1.5.14}$$

Here vector $\vec{R}_{i+1,j}$ is the distance vector between centers of joints j and i+1. Applying the methods presented in Subsections 1.3.1 and 1.3.2, vectors $\vec{e}_j$ and $\vec{R}_{i+1,j}$ can be expressed with respect to any coordinate frame. Therefore, the linear velocity $\vec{v}_i$ can be obtained with respect to the reference frame, as well as to any link coordinate system.

The linear velocity of the manipulator tip (end-effector) is obtained from (1.5.14) for i=n

$$\vec{v}_n = \sum_{j=1}^{n} [(\vec{e}_j \times \vec{R}_{n+1,j}) (1-\xi_j) + \vec{e}_j \xi_j] \dot{q}_j \tag{1.5.15}$$

This can be presented in the following matrix form

$$\vec{v}_n = J_c \dot{q}, \qquad J_c \in R^{3 \times n} \tag{1.5.16}$$

where

$$J_c = [(\vec{e}_1 \times \vec{R}_{n+1,1})(1-\xi_1) + \vec{e}_1 \xi_1 \; \vdots \; \cdots \; \vdots \; (\vec{e}_n \times \vec{R}_{n+1,n})(1-\xi_n) + \vec{e}_n \xi_n] \tag{1.5.17}$$

Those columns of matrix J_c which correspond to the revolute joints represent vectors $\vec{e}_j \times \vec{R}_{n+1,j}$, while the columns which correspond to the sliding joints represent vectors $\vec{e}_j$ (unit vectors of joint axes). Vector $\vec{R}_{n+1,j}$ (see Fig. 1.21) is the distance vector from the center of

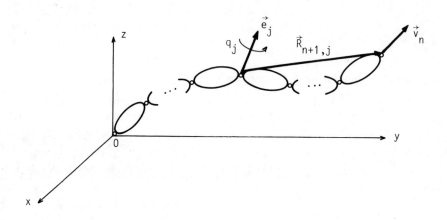

Fig. 1.21. Linear velocity of the manipulator hand

joint j to the manipulator tip. If vectors $\vec{e}_j$ and $\vec{R}_{n+1,j}$, $j=1,\ldots,n$ are expressed with respect to the base reference frame, then relations (1.5.16) and (1.5.17) yield the Cartesian coordinate velocity of the manipulator hand $^O\vec{v}_n = [\dot{x} \; \dot{y} \; \dot{z}]^T$ with respect to the base reference frame, given a vector of joint rates $\dot{q}$. If these vectors are expressed with respect to the coordinate frame attached to the last link, then we obtain the linear hand velocity with respect to the end-effector coordinate frame.

1.5.2. Jacobian matrices

We have seen in Sections 1.3 and 1.4 that we can determine manipulator hand position and orientation, given a joint coordinate vector q. The vectorial function

$$x_e = f(q) \qquad (1.5.18)$$

is thus determined, where $q \in Q \subset R^n$, $x_e \in X_e \subset R^m$, $f: R^n \to R^m$. The nonlinear function f is continuous and differentiable for $\forall q \in Q$. Note that this function does not map the set X_e into the set Q uniquely, but Q is uniquely mapped into X_e by this function.

By differentiating Equation (1.5.18) with respect to time we obtain

$$\frac{dx_e}{dt} = \frac{\partial f(q)}{\partial q} \frac{dq}{dt}$$

or

$$\dot{x}_e = J\dot{q} \qquad (1.5.19)$$

where $J = \frac{\partial f(q)}{\partial q} \in R^{m \times n}$ is the Jacobian matrix of partial derivatives. This matrix relates between joint rates and external coordinate veloci- ties. The elements of matrix J depend on joint coordinate vector q. The computation of this matrix is very important in manipulator control, especially for rate control, inverse kinematic problem solution, as well as for treating external forces which act on the manipulator grip- per. For this reason, we will consider computation of the Jacobian in detail in the following text.

As the vector of external coordinates may be adopted in various manners (see Section 1.2), the computation of this matrix differs, too.

We will here consider the computation of Jacobian matrices for the case where the dimension of vector x_e is m=6, this being the most general case in practical, industrial manipulation, i.e. we will assume that the hand position and orientation are described by three coordinates each. Then Equation (1.5.19) can be presented as

$$\dot{x}_e = \begin{bmatrix} \dot{x}_{eI} \\ ----- \\ \dot{x}_{eII} \end{bmatrix} = J\dot{q} = \begin{bmatrix} J_I \\ ----- \\ J_{II} \end{bmatrix} \dot{q} \qquad (1.5.20)$$

where $x_{eI} \in R^3$ is the position vector, $x_{eII} \in R^3$ - orientation vector (see Equations (1.2.2) - (1.2.6)), $J_I \in R^{3 \times n}$ and $J_{II} \in R^{3 \times n}$ are corresponding Jacobian submatrices. We will first consider the computation of the

Jacobian submatrices $J_I \in R^{3 \times n}$ for various position vectors:

^oJ_c - for the Cartesian coordinates in the reference frame

^nJ_c - the Cartesian coordinates in the hand coordinate frame

J_s - the spherical coordinates

J_{cy} - cylindrical coordinates.

We will then derive expressions for submatrices $J_{II} \in R^{3 \times n}$ for various orientation specifications:

$^oJ_\omega$ - for the end-effector angular velocities expressed in the base reference frame

$^nJ_\omega$ - the end-effector angular velocities expressed in the coordinate system of the last link

J_E - Euler angles

J_{Ep} - Euler parameters.

Jacobian matrix for Cartesian coordinates of the base reference system

Let us consider the case when the manipulator hand position is described by Cartesian coordinates of the reference coordinate frame Oxyz attached to manipulator base, i.e.

$$x_{eI} = [x \ y \ z]^T$$

In the previous subsection we have derived expressions (1.5.16) and (1.5.17) which relate between the translational velocity of the manipulator tip and the joint rates.

$$^o\vec{v}_n = [\dot{x} \ \dot{y} \ \dot{z}]^T = {}^oJ_c \dot{q} \qquad (1.5.21)$$

where $^oJ_c \in R^{3 \times n}$ is the Cartesian Jacobian matrix with respect to the base reference system

$$^oJ_c = [\, ({}^o\vec{e}_1 \times {}^o\vec{R}_{n+1,1})(1-\xi_1) + {}^o\vec{e}_1\xi_1 \mid \cdots \mid ({}^o\vec{e}_n \times {}^o\vec{R}_{n+1,n})(1-\xi_n) + {}^o\vec{e}_n\xi_n]$$
$$(1.5.22)$$

Here $\overset{o}{\vec{e}}_i$ denotes the unit vectors of the joint axes, expressed in the reference frame, and $\overset{o}{\vec{R}}_{n+1,i}$ the position vectors from centers of joints i, i=1,...,n to the manipulator tip, expressed in the reference frame (Fig. 1.21). For an alternative approach to deriving the Jacobian matrix (1.5.22) see [1].

Equation (1.5.22) can be used either for the numerical computation of the elements of $\overset{o}{J}_c$, or for deriving symbolic expressions for these elements in a form of the trigonometric functions of joint coordinates and the kinematic parameters. Symbolic kinematic model generation for any specific manipulator, by means of a computer, will be discussed in detail in Chapter 2.

Here, we will consider the numerical computation of the Jacobian matrix, since it is still frequently used in manipulator control. The predominant concern is the computational complexity. This depends on several factors: how the link coordinate frames are attached to the links (Rodrigues formula approach or Denavit-Hartenberg kinematic notation), which type of recursivness is adopted (recursions from base toward the end-effector or inversely), whether Jacobian $\overset{o}{J}_c$ with respect to the reference frame, or Jacobian $\overset{n}{J}_c$ with respect to the manipulator end-effector is considered (this matrix will be discussed in the text to follow). An analysis of computational efficiency of several methods was given in [14]. One method for computing the matrix $\overset{o}{J}_c$ (i.e. the reference point is the manipulator end-effector tip and the translational velocity of this point with respect to the base reference system is considered), is to use recursive relations (1.3.11), (1.3.17), i.e. according to

$$\overset{o}{A}_i = \overset{o}{A}_{i-1}\,\overset{i-1}{A}_i, \qquad i=2,\ldots,n$$

$$\overset{o}{\vec{e}}_i = \overset{o}{A}_{i-1}\,\overset{i-1}{\vec{e}}_i$$

$$\overset{o}{\vec{R}}_{i,o} = \overset{o}{\vec{R}}_{i-1,o} + \overset{o}{A}_{i-1}\,\overset{i-1}{\vec{R}}_{i,i-1}, \qquad i=2,\ldots,n \qquad (1.5.23)$$

$$\overset{o}{\vec{R}}_{n+1,i} = \overset{o}{\vec{R}}_{n+1,o} - \overset{o}{\vec{R}}_{i,o}$$

$$\overset{o}{j}_o^{(i)} = \begin{cases} \overset{o}{\vec{e}}_i \times \overset{o}{\vec{R}}_{n+1,i} & \text{- for revolute joints} \\ \overset{o}{\vec{e}}_i & \text{- for sliding joints} \end{cases}$$

$$\overset{o}{J}_c = [\overset{o}{j}_c^{(1)} \mid \cdots \mid \overset{o}{j}_c^{(n)}] \in R^{3 \times n}$$

The relations are the most appropriate for Rodrigues formula approach, where joint axes ${}^{i-1}\vec{e}_i$ are arbitrary unit vectors with respect to co-ordinate systems (i-1) (Fig. 1.22).

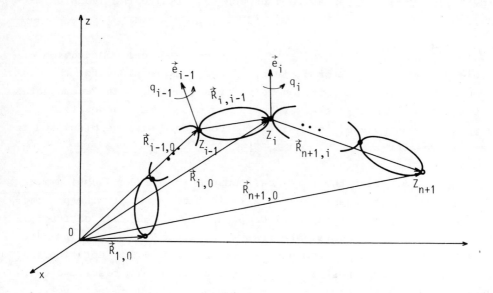

Fig. 1.22. Vectors relevant for the Jacobian computation

However, the joint axes $\vec{e}_i$ are most frequently colinear with one of the axes of the link coordinate system. In this case matrix-vector product ${}^{O}A_{i-1}\,{}^{i-1}\vec{e}_i$ requires no floating-point operations. The computation of the vector cross product can be similarly eliminated, as will be described in the text to follow, for Denavit-Hartenberg kinematic notation.

In Denavit-Hartenberg kinematic notation the joint axes coincide with $\vec{z}_{i-1}$ axes of link coordinate systems, so that

$$^{i-1}\vec{e}_i = \vec{z}_{i-1} = [0\ 0\ 1]^T$$

holds. This can be easily utilized to avoid the vector cross product computation, if the distance vector between the joint and manipulator tip is expressed with respect to system (i-1), too, and subsequently the cross product (which requires no mathematical operations) is transformed into the reference frame premultiplying by ${}^{O}A_{i-1}$. Namely, if we introduce the following notation for the parts of the homogeneous transformation matrix

$$^{i-1}T_n = \left[\begin{array}{c|c} ^{i-1}A_n & ^{i-1}\vec{P}_n \\ \hline 0\ 0\ 0 & 1 \end{array}\right] \tag{1.5.24}$$

where $^{i-1}\vec{P}_n \in R^3$ is the distance vector from the center of joint i (origin of coordinate frame i-1) to the manipulator tip (origin of coordinate system n), then the vector cross product

$$^{i-1}\vec{e}_i \times {}^{i-1}\vec{R}_{n,i} = \vec{z}_{i-1} \times {}^{i-1}\vec{P}_n = [0\ 0\ 1]^T \times [{}^{i-1}P_{nx}\ {}^{i-1}P_{ny}\ {}^{i-1}P_{nz}]^T =$$

$$[-^{i-1}P_{ny}\ {}^{i-1}P_{nx}\ 0]^T \tag{1.5.25}$$

requires no multiplications or additions. Therefore, it is convenient to compute the columns of Jacobian matrix ^OJ_c in the following way

$$^OA_i = {}^OA_{i-1}\ {}^{i-1}A_i, \qquad i=2,\dots,n$$

$$^{i-1}\vec{P}_n = {}^{i-1}T_i \left[\begin{array}{c} ^i\vec{P}_n \\ \hline 1 \end{array}\right] = {}^{i-1}A_i\ {}^i\vec{P}_n + {}^{i-1}\vec{P}_i, \qquad i=n,\dots,1$$

$$^OJ_c(i) = {}^OA_{i-1} \left[\begin{array}{c} -^{i-1}P_{ny} \\ -^{i-1}P_{nx} \\ 0 \end{array}\right], \qquad \text{for revolute joints} \tag{2.5.26}$$

$$^OJ_c(i) = {}^OA_{i-1} \left[\begin{array}{c} 0 \\ 0 \\ 1 \end{array}\right], \qquad \text{for sliding joints}$$

We can see that in (1.5.26) transformation (rotation) matrices ^OA_i are computed by forward recursions, while position vectors are computed by backward recursions (Fig. 1.23).

The above discussion reffers only to the computation of the Jacobian which relates between the joint rates and the translational velocities of the manipulator hand in the base reference system.

44

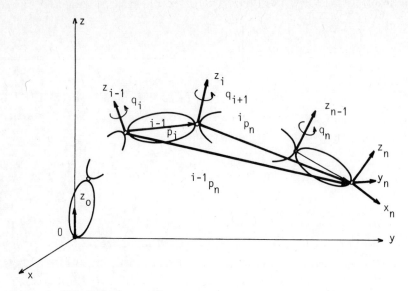

Fig. 1.23. Vectors relevant for the Jacobian computation
in Denavit-Hartenberg notation

*Jacobian matrix for Cartesian coordinates of
the hand coordinate frame*

Let us now consider the relationship between the joint coordinate velocities and the translational velocities of the manipulator hand, expressed with respect to the coordinate frame assigned to the last link

$$^n\vec{v}_n = [^n\dot{x} \; ^n\dot{y} \; ^n\dot{z}]^T = {}^nJ_c\dot{q} \tag{1.5.27}$$

where $^nJ_c \in R^{3 \times n}$. This matrix is obtained directly from Equation (1.5.17),
if the joint axes' vectors and the position vectors are expressed with
respect to the coordinate frame of link n. Thus we have

$$^nJ_c = [\,(^n\vec{e}_1 \times {}^n\vec{R}_{n+1,1})\,(1-\xi_1) + {}^n\vec{e}_1\xi_1 \;|\; \cdots \;|\; (^n\vec{e}_n \times {}^n\vec{R}_{n+1,n})\,(1-\xi_n) + {}^n\vec{e}_n\xi_n\,]$$

$$\tag{1.5.28}$$

where $^n\vec{e}_i$ denotes the unit vector of the joint axis of link i expressed
in frame n, $^n\vec{R}_{n+1,i}$ is the position vector from the center of joint i
to the manipulator tip, expressed in frame n, $\xi_i = 0$ if joint i is a rotational one and $\xi_i = 1$ of joint i is prismatic.

Equation (1.5.28) can be used either for generating the symbolic form of matrix $^{n}J_{c}$ for a given manipulator (see Chapter 2), or for numerical computation of this matrix. An efficient method for computing this matrix, is by the following set of recursive relations

$$^{i}A_{n} = {}^{i}A_{i+1}\,{}^{i+1}A_{n}, \qquad i=n-1,\ldots,1,0$$

$$^{i}R_{n+1,i} = {}^{i}R_{i+1,i} + {}^{i}A_{i+1}\,{}^{i+1}\vec{R}_{n+1,i+1}, \qquad i=n-1,\ldots,1,0$$

$$^{n}j_{c}(i) = {}^{i}A_{n}^{T}({}^{i}\vec{e}_{i} \times {}^{i}\vec{R}_{n+1,i}) \quad - \text{ for revolute joints} \qquad (1.5.29)$$

$$^{n}j_{c}(i) = {}^{i}A_{n}^{T}\,{}^{i}\vec{e}_{i} \quad - \text{ for sliding joints}$$

$$^{n}J_{c} = [\,{}^{n}j_{c}(1)\,|\,\cdots\,|\,{}^{n}j_{c}(n)\,] \in R^{3 \times n}$$

We see that the transformation matrices and position vectors are calculated by backward recursive relations, i.e. for $i=n-1,\ldots,1,0$. If the joint axis e_{i} is colinear with some of the unit vectors of the coordinate system i, the above vector cross-product is simplified and requires no floating point operations.

If homogeneous transformations are used as a kinematic modelling technique, the first two equations of (1.5.29) are replaced by

$$^{i-1}T_{n} = {}^{i-1}T_{i}\,{}^{i}T_{n}, \qquad i=n,\ldots,1 \qquad (1.5.30)$$

If the link coordinate frames are assigned to the links according to Denavit-Hartenberg method, the vector cross product in (1.5.29) becomes $\vec{z}_{i-1} \times {}^{i-1}\vec{P}_{n} = [-{}^{i-1}P_{ny} \quad {}^{i-1}P_{nx} \quad 0]^{T}$ (similarly to (1.5.25)). So the columns of $^{n}J_{c}$ are

$$^{n}j_{c}(i) = {}^{i-1}A_{n}^{T}\begin{bmatrix} -{}^{i-1}P_{ny} \\ -{}^{i-1}P_{nx} \\ 0 \end{bmatrix}, \quad \text{ for revolute joints} \qquad (1.5.31)$$

$$^{n}j_{c}(i) = {}^{i-1}A_{n}^{T}\begin{bmatrix} 0 \\ 0 \\ 1 \end{bmatrix}, \quad \text{ for prismatic joints} \qquad (1.5.32)$$

Let us remember, once again, that $^{i-1}A_{n}$ and $^{i-1}\vec{P}_{n}$ are the submatrices of $^{i-1}T_{n}$ according to (1.5.24).

Jacobian matrices ^nJ_c and ^oJ_c are related by

$$^oJ_c = {^oA_n}\,{^nJ_c} \tag{1.5.33}$$

This is the result of the definitions of these matrices (Eqs. (1.5.21) and (1.5.27)) and the relation between the linear hand velocity expressed with respect to system n and the base reference system

$$^o\vec{v}_n = {^oA_n}\,{^n\vec{v}_n} \tag{1.5.34}$$

Jacobian matrix for spherical coordinates

Let us now consider a case in which the manipulator hand position is specified by the spherical coordinates r, β and α of the fixed reference frame, attached to the manipulator base (Fig. 1.3), i.e.

$$x_{eI} = [r\ \beta\ \alpha]^T$$

Let us determine the matrix relating between the velocities of these coordinates and the joint rates

$$\dot{x}_{eI} = [\dot{r}\ \dot{\beta}\ \dot{\alpha}]^T = J_s\dot{q}$$

where $J_s \epsilon R^{3\times n}$. This matrix can be obtained by indirect means by utilizing the Jacobian matrix ^oJ_c corresponding to Cartesian coordinates, as follows: by differentiating equations (1.3.29) - (1.3.31) which relate between the spherical and Cartesian coordinates, we obtain

$$
\begin{bmatrix} \dot{x} \\ \dot{y} \\ \dot{z} \end{bmatrix} =
\begin{bmatrix}
\cos\alpha\sin\beta & r\cos\alpha\cos\beta & -r\sin\alpha\sin\beta \\
\sin\alpha\sin\beta & r\sin\alpha\cos\beta & r\cos\alpha\sin\beta \\
\cos\beta & -r\sin\beta & 0
\end{bmatrix}
\begin{bmatrix} \dot{r} \\ \dot{\beta} \\ \dot{\alpha} \end{bmatrix} \tag{1.5.35}
$$

On the other hand, we have the well known relation $[\dot{x}\ \dot{y}\ \dot{z}]^T = {^oJ_c}\dot{q}$. Substituting this relation into (1.5.35) and premultiplying by the matrix inverse, we get

$$
\begin{bmatrix} \dot{r} \\ \dot{\beta} \\ \dot{\alpha} \end{bmatrix} = J_s\dot{q} =
\begin{bmatrix}
\cos\alpha\sin\beta & \sin\alpha\sin\beta & \cos\beta \\
\dfrac{1}{r}\cos\alpha\cos\beta & \dfrac{1}{r}\sin\alpha\cos\beta & -\dfrac{1}{r}\sin\beta \\
\dfrac{-\sin\alpha}{r\sin\beta} & \dfrac{\cos\alpha}{r\sin\beta} & 0
\end{bmatrix}
{^oJ_c}\dot{q} \tag{1.5.36}
$$

Thus Jacobian matrix J_s corresponding to the spherical coordinates is expressed by means of Cartesian Jacobian matrix ^oJ_c, given by (1.5.22).

However, Equation (1.5.36) shows that some matrix elements diverge for small values of r and $\sin\beta$. Therefore, the use of matrix J_s is unsuitable for motion generation in the space of spherical coordinates. Instead of utilizing matrix J_s, motion generation can be carried out as follows: given the velocities $\dot{r}$, $\dot{\beta}$ and $\dot{\alpha}$, the corresponding Cartesian velocities are calculated by using Equation (1.5.35). The problem of motion synthesis in spherical coordinates is thus reduced to Cartesian motion generation (the inverse kinematic problem is solved using Jacobian ^oJ_c instead of J_s).

Jacobian matrix for cylindrical coordinates

The Jacobian matrix relating between the cylindrical coordinate velocities $\dot{r}$, $\dot{\alpha}$, $\dot{z}$ and the joint velocities is obtained by the same method as for the spherical coordinates. By differentiating Equations (1.3.36)-(1.3.38) we obtain

$$\begin{bmatrix} \dot{x} \\ \dot{y} \\ \dot{z} \end{bmatrix} = \begin{bmatrix} \cos\alpha & -r\sin\alpha & 0 \\ \sin\alpha & r\cos\alpha & 0 \\ 0 & 0 & 1 \end{bmatrix} \begin{bmatrix} \dot{r} \\ \dot{\alpha} \\ \dot{z} \end{bmatrix} \tag{1.5.37}$$

Taking that $[\dot{x}\ \dot{y}\ \dot{z}]^T = {}^oJ_c\dot{q}$, we get

$$\begin{bmatrix} \dot{r} \\ \dot{\alpha} \\ \dot{z} \end{bmatrix} = J_{cy}\dot{q} = \begin{bmatrix} \cos\alpha & \sin\alpha & 0 \\ -\frac{1}{r}\sin\alpha & \frac{1}{r}\cos\alpha & 0 \\ 0 & 0 & 1 \end{bmatrix} {}^oJ_c\dot{q} \tag{1.5.38}$$

where ^oJ_c is the Cartesian Jacobian matrix given by (1.5.22).

As with the spherical coordinates, some elements of the matrix in (1.5.38) diverge for small values of coordinate r. A direct consequence of this is that matrix J_{cy} is inconvenient for motion synthesis in the space of cylindrical coordinates. Instead of that, given the velocities $\dot{r}$, $\dot{\alpha}$ and $\dot{z}$ one should compute the corresponding Cartesian velocities $\dot{x}$, $\dot{y}$, $\dot{z}$ by means of (1.5.37), and then carry out the synthesis as for the Cartesian coordinates (using ^oJ_c instead of J_{cy}).

Jacobian matrix relating between the end-effector
angular velocities and joint velocities

Let us assume that the rate of changing the hand orientation is specified by the angular velocity of the last link, expressed with respect
to the base reference frame, i.e.

$$\dot{x}_{eII} = [{}^{O}\omega_{nx} \quad {}^{O}\omega_{ny} \quad {}^{O}\omega_{nz}]^{T} \tag{1.5.39}$$

The Jacobian matrix ${}^{O}J_{\omega} \in R^{3 \times n}$, defined by

$$\dot{x}_{eII} = [{}^{O}\omega_{nx} \quad {}^{O}\omega_{ny} \quad {}^{O}\omega_{nz}]^{T} = {}^{O}J_{\omega}\dot{q} \tag{1.5.40}$$

is yet to be determined. On the other hand, it was shown in Subsection
1.5.1, that the matrix relating between the angular velocity of link n
and the joint rates (Eq. (1.5.7)), has the following form

$$
{}^{O}J_{\omega} = [{}^{O}\vec{e}_{1}(1-\xi_{1}) \mid \cdots \mid {}^{O}\vec{e}_{n}(1-\xi_{n})] \tag{1.5.41}
$$

where ${}^{O}\vec{e}_{i}$ denotes the unit vectors of the joint axes in the base rere-
rence frame, $\xi_{i}=0$ for a rotational joint and $\xi_{i}=1$ for a sliding joint.

In Denavit-Hartenberg kinematic notation, the relation ${}^{i-1}\vec{e}_{i} = \vec{z}_{i-1} =$
$= [0 \ 0 \ 1]^{T}$ holds so that

$$
{}^{O}\vec{e}_{i} = {}^{O}A_{i-1} \begin{bmatrix} 0 \\ 0 \\ 1 \end{bmatrix}, \qquad i=1,\ldots,n \tag{1.5.42}
$$

This implies that the columns of matrix ${}^{O}J_{\omega}$ are either zero columns,
or coincide with the third columns of the rotation matrices ${}^{O}A_{i-1}$.

If the rate of changing the hand orientation is specified by the angular velocity of the last link, expressed with respect to the coordinate
system of link n

$$\dot{x}_{eII} = [{}^{n}\omega_{nx} \quad {}^{n}\omega_{ny} \quad {}^{n}\omega_{nz}]^{T} \tag{1.5.43}$$

the Jacobian matrix ${}^{n}J_{\omega} \in R^{3 \times n}$ is defined by

$$\dot{x}_{eII} = [{}^{n}\omega_{nx} \quad {}^{n}\omega_{ny} \quad {}^{n}\omega_{nz}]^{T} = {}^{n}J_{\omega}\dot{q} \tag{1.5.44}$$

Matrix $^nJ_\omega$ is also obtained from Eq. (1.5.7) as

$$^nJ_\omega = [\,^n\vec{e}_1(1-\xi_1) \mid \cdots \mid \,^n\vec{e}_n(1-\xi_n)\,] \qquad (1.5.45)$$

where the unit vectors of the joint axes are expressed in the coordinate frame of link n.

For Denavit-Hartenberg kinematic notation, vectors $^n\vec{e}_i$ are given by

$$^n\vec{e}_i = \,^{i-1}A_n^T \begin{bmatrix} 0 \\ 0 \\ 1 \end{bmatrix}, \qquad i=1,\ldots,n \qquad (1.5.46)$$

so that the columns of matrix $^nJ_\omega$ are either zero columns, or represent the third row of the transformation matrix $^{i-1}A_n$.

It is evident from equations (1.5.41) and (1.5.45) that the relationship between matrices $^oJ_\omega$ and $^nJ_\omega$ is given by

$$^oJ_\omega = \,^oA_n\,^nJ_\omega \qquad (1.5.47)$$

Matrix $^nJ_\omega$ is directly utilized in manipulator motion generation, when the change of gripper orientation is specified by angular velocities about the gripper axes. This is convenient, for example, in teaching the manipulator by means of a teaching box, when precise adjustments in manipulator orientation are made. On the other hand matrix $^oJ_\omega$ is used for the manipulator transfer movements, when Euler angle velocities are specified and then transformed into the corresponding angular velocities $^o\omega_{nx}$, $^o\omega_{ny}$, $^o\omega_{nz}$ with respect to the base reference frame.

Jacobian matrix for Euler angles

Let us now consider the computation of the lower Jacobian submatrix J_{II} (Eq. (1.5.20)), for the case when end-effector orientation is specified by Euler angles yaw, pitch and roll (Fig. 1.18)

$$x_{eII} = [\psi \;\; \theta \;\; \varphi]^T$$

The matrix J_E, relating between the rate of these coordinates and joint velocities, is to be determined

$$\dot{x}_{eII} = [\dot{\psi} \ \dot{\theta} \ \dot{\phi}]^T = J_E \dot{q} \qquad (1.5.48)$$

The relationship between angular velocities of a rigid body and Euler angle velocities is well known from the kinematics of a rigid body

$$\begin{bmatrix} {}^0\omega_{nx} \\ {}^0\omega_{ny} \\ {}^0\omega_{nz} \end{bmatrix} = \begin{bmatrix} 0 & -\sin\psi & \cos\theta\cos\psi \\ 0 & \cos\psi & \cos\theta\sin\psi \\ 1 & 0 & -\sin\theta \end{bmatrix} \begin{bmatrix} \dot{\psi} \\ \dot{\theta} \\ \dot{\phi} \end{bmatrix} \qquad (1.5.49)$$

On the other hand, the relation $[{}^0\omega_{nx} \ {}^0\omega_{ny} \ {}^0\omega_{nz}]^T = {}^0J_\omega \dot{q}$ is valid. Substituting this relation into (1.5.49) and premultiplying by the matrix inverse, ve obtain

$$\begin{bmatrix} \dot{\psi} \\ \dot{\theta} \\ \dot{\phi} \end{bmatrix} = J_E \dot{q} = \begin{bmatrix} \cos\psi \, \text{tg}\theta & \sin\psi \, \text{tg}\theta & 1 \\ -\sin\psi & \cos\psi & 0 \\ \dfrac{\cos\psi}{\cos\theta} & \dfrac{\sin\psi}{\cos\theta} & 0 \end{bmatrix} {}^0J_\omega \dot{q} \qquad (1.5.50)$$

where ${}^0J_\omega$ is given by equation (1.5.41).

Matrix J_E relates directly between the Euler angle rates and joint velocities. Unfortunately, this matrix is very unsuitable for manipulator motion generation, since $\cos\theta$ figures in the denominator of some matrix elements. Therefore, when the hand orientation is specified by Euler angles, motion synthesis is not carried out by using matrix J_E, but by the following method: given the Euler angle rates $\dot{\psi}$, $\dot{\theta}$ and $\dot{\phi}$, the corresponding angular velocities of the last link are computed using (1.5.49), and then matrix ${}^0J_\omega$ is used for the solution of the inverse kinematic problem.

When there are less than 6 degrees of freedom manipulator hand orientation (Euler angles) is not completely controllable. Some of Euler angle rates can be imposed along the trajectory, depending on manipulator configuration and the number of degrees of freedom ($3<n<6$). The remaining angles can not be changed, so that their velocities have to be kept at zero. However, according to (1.5.49), all the three angular velocities vary, although some of the $\dot{\psi}$, $\dot{\theta}$ or $\dot{\phi}$ velocities are zero. In that case if we want to avoid the use of a submatrix of J_E in motion synthesis, we can solve the overdetermined system

$$\dot{x}_e = \begin{bmatrix} \dot{x} \\ \dot{y} \\ \dot{z} \\ {}^o\omega_{nx} \\ {}^o\omega_{ny} \\ {}^o\omega_{nz} \end{bmatrix} = J\dot{q} = \begin{bmatrix} {}^oJ_c \\ \hline {}^oJ_\omega \end{bmatrix} \dot{q}, \qquad J \in R^{6 \times n}, \qquad 3 < n < 6 \qquad (1.5.51)$$

The least-square solution of system (1.5.51) is

$$\dot{q} = (J^T J)^{-1} J^T \dot{x}_e \qquad (1.5.52)$$

where $\dot{q} \in R^n$, $J \in R^{6 \times n}$, $n < 6$, $\dot{x}_e \in R^6$.

Jacobian matrix for Euler parameters

In the case when hand orientation is specified by Euler parameters (see Subsection 1.4.2)

$$x_{eII} = [\lambda_1 \ \lambda_2 \ \lambda_3]^T$$

The Jacobian matrix $J_{E_p} \in R^{3 \times n}$ relates between the rate of changing Euler parameters and the joint rates

$$\dot{x}_{eII} = [\dot{\lambda}_1 \ \dot{\lambda}_2 \ \dot{\lambda}_3]^T = J_{E_p} \dot{q}$$

This matrix is shown to have the form [4, 15]

$$J_{E_p} = \frac{1}{2} \begin{bmatrix} \lambda_o & \lambda_3 & -\lambda_2 \\ -\lambda_3 & \lambda_o & \lambda_1 \\ \lambda_2 & -\lambda_1 & \lambda_o \end{bmatrix} {}^oJ_\omega$$

where matrix ${}^oJ_\omega$ is given by (1.5.41).

1.6. Summary

In this chapter we have presented main mathematical relations and definitions describing the manipulator end-effector position and orientation and it''s velocities. Two methods for kinematic modelling have been outlined. The first method makes use of Rodrigues formula describing a finite rotation of a rigid body about a fixed axis. The second approach uses the homogenous transformations with Denavit-Hartenberg kinematic notation of open kinematic chains. In the first method link coordinate frames are, from the kinematic point of view, centered and oriented arbitrarily with respect to the joint centers and axes. In the second approach the link coordinate systems are defined by the disposition of joint axes. If in the Rodrigues formula approach the link coordinate frames were adopted to coincide with Denavit-Hartenberg link frames, then these two methods would be practically equal with respect to the computational complexity.

Manipulator hand position is usually specified with respect to a Cartesian reference frame attached to the manipulator base. The specification of the manipulator end-effector position by means of spherical and cylindrical coordinates, together with Cartesian coordinates, has been described in this chapter.

Euler angles and Euler parameters were used in this chapter in order to describe end-effector orientation.

Jacobian matrices relating between linear and angular velocities of the end-effector and joint coordinates have been derived. Jacobian matrices with respect to the base reference system and the hand coordinate frame have been determined. Efficient methods have been presented for computing these matrices.

Chapter 2
Computer-aided Generation of Kinematic Equations in Symbolic Form

2.1. Introduction

Symbolic kinematic equations of robotic manipulators play an important role in contemporary robot control. The use of symbolic equations instead of numeric ones, speeds up the computation of the control signals necessary to guide a manipulator along a desired path. On the other hand, on-line motion generation, by means of low-cost microcomputers, becomes an important feature of modern industrial robots. In order to achieve this feature, the least computationally expensive algorithms should be applied. Such algorithms certainly make use of the symbolic kinematic model. The kinematic equations in symbolic form are equations which describe manipulator hand position and orientation as explicit trigonometric functions of the joint coordinates. These equations can be obtained by hand using the equations derived in the preceding chapter. The same is true in the case of the derivation of the symbolic Jacobian matrix. However, this tiresome process is subjected to human errors. Besides, once the symbolic equations have been obtained, a computer program must be written to compute these kinematic variables optimally, i.e. with the minimal number of multiplications and additions. In this chapter we will do this automatically, by means of a computer.

A method will be presented to enable the automatic, computer-aided generation of the kinematic model of an n degree-of-freedom manipulator in symbolic form. The output of the algorithm is a computer program for evaluating manipulator hand position and the transformation matrix between the end-effector coordinate frame and the base, reference frame as well as the Jacobian matrix. The computation is carried out optimally in the sense of minimal number of floating-point multiplications and additions. Thus, the main objectives to be attained by this procedure can be summarized as:

1. to avoid numeric computation of the kinematic equations and Jacobian matrix in on-line motion generation;

2. to avoid manual derivation of the kinematic model;

3. to obtain programs for computing this model optimally with respect

to the number of floating-point operations.

The problem of computer-aided symbolic kinematic model generation has already been considered in [16-18, 69], as the first phase in dynamic model generation. This approach is based on the application of Rodrigues' formula, presented in Subsection 1.3.1. This method, however, yields a numeric-symbolic model, but not a pure symbolic model, where the kinematic parameters are treated as numeric constants instead of variables. If any kinematic parameter is changed, the model should be regenerated. This procedure, however, does not give kinematic equations in the form having the minimal number of floating-point multiplications and additions. Computational redundancy increases with the number of degrees of freedom of the mechanism.

The algorithm for computer-aided generation of the kinematic model, which will be described in this chapter, is based on Denavit-Hartenberg's kinematic notation. It makes use of either backward or forward recursive symbolic relations in computing elements of the homogeneous transformation matrices. The direction of the recursiveness depends on whether or not the Jacobian is to be formed with respect to the hand or the base reference system. The recursive relations are symbolic. They are derived starting from the ordinary recursive relations presented in the Subsections 1.3.2 and 1.5.2. The problem of compressing the trigonometric expressions is avoided either by combining backward and forward relations, or carried out analytically on the general symbolic expressions.

The algorithm for symbolic model generation has been achieved in the form of a program package. Its input variables are Denavit-Hartenberg's kinematic parameters of the manipulator. On output it gives a source program for computing homogeneous transformation matrix $^{O}T_n$ between the coordinate system of the final link and the base reference frame, as well as the Jacobian matrix ^{O}J or ^{n}J relating between Cartesian and angular end-effector velocities (expressed either in the base or hand coordinate system) and the joint rates.

In Section 2.2 we will be concerned with the generation of matrix $^{O}T_n$. Once the matrix $^{O}T_n$ is computed, given a vector of joint coordinates q, all other types of external coordinates (spherical, cylindrical, Euler-angles) can be readily obtained, as was shown in the previous chapter. Therefore, in the following text we will assume that the solution to the direct kinematic problem is equivalent to the computing of homogeneous matrix $^{O}T_n$.

In order to describe the derivation of symbolic recursive relations for solving the direct kinematic problem, we will first illustrate the manual use of homogeneous transformations. This example, given for the manipulator UMS-3B, will later be compared with the computer-generated set of equations. Backward recursive symbolic relations for the elements of homogeneous transformation matrices $^{i-1}T_n$, i=n,...,1, will be derived in Subsection 2.2.3. They comprise both revolute and prismatic joints. Forward recursive symbolic relations, yielding elements of matrices $^{O}T_i$, i=1,...,n are derived in Subsection 2.2.4. The application of these relations is illustrated by the example of manipulator UMS-3B. As is to be expected, revolute joints with parallel joint axes require special treatment if the minimal number of floating-point operations is demanded. Backward and forward recursive symbolic relations for revolute joints with parallel axes will be developed and illustrated in Subsection 2.2.5.

Section 2.3 deals with the generation of Jacobian matrix ^{n}J which relates between the linear and angular velocities of the manipulator hand, expressed in the hand coordinate frame, and joint velocities. We will term this matrix as the Jacobian with respect to the hand coordinate frame. We will first note down the Jacobian for the manipulator UMS-3B. Then, we will point out the problem of condensing the trigonometric expressions, which emerges in forming the Jacobian. The problem will be solved analytically, so that the symbolic relations derived for the Jacobian elements guarantee practically minimal computational complexity. Special relations are derived for those Jacobian columns which correspond to joints with parallel axes.

The Jacobian matrix ^{O}J which relates between linear and angular hand velocities, expressed in the base coordinate system, and joint velocities, will be considered in Section 2.4. The compression of trigonometric expressions is avoided by combining the backward and forward recursive symbolic relations derived in Subsections 2.2.3 and 2.2.4. The Jacobian columns which correspond to revolute joints with parallel joint axes are formed using special symbolic relations.

Section 2.5 describes the organization of the software package for generating the symbolic model. The main idea is that, by using the previously derived symbolic recursive relations complex variables which require floating-point multiplications or additions are entered into an output file, while the simple variables are stored in the computer memory. Thus, the output file contains the instructions necessary for computing matrices $^{0}T_{n}$ and ^{0}J or ^{n}J.

The last section deals with numerical aspects of the results. The numbers of floating-point multiplications and additions needed, and the program execution times for several types of industrial robots are presented here. Computational complexity is also compared with the corresponding numeric model computation. The computer program for manipulator UMS-3B, obtained by the software package is also given.

2.2. Symbolic Kinematic Equations

2.2.1. Backward and forward recursive relations

This kinematic analysis will be based on Denavit-Hartenberg's kinematic notation, presented in Subsection 1.3.2 in detail. For the sake of clearness, we will here repeat the expressions for homogeneous transformation matrices $^{i-1}T_{i}$ between two adjacent link coordinate frames.

If joint i is a revolute one, and link coordinate frames are assigned to the links according to the procedure presented in Subsection 1.3.2 (Fig. 1.16), the homogeneous transformation between systems i and i-1, is given by

$$^{i-1}T_{i} = \begin{bmatrix} \cos q_i & -\sin q_i \cos \alpha_i & \sin q_i \sin \alpha_i & a_i \cos q_i \\ \sin q_i & \cos q_i \cos \alpha_i & -\cos q_i \sin \alpha_i & a_i \sin q_i \\ 0 & \sin \alpha_i & \cos \alpha_i & d_i \\ 0 & 0 & 0 & 1 \end{bmatrix} \qquad (2.2.1)$$

where: q_i - is the rotation angle in joint i, measured between axes x_{i-1} and x_i (Fig. 1.16);

α_i - the twist angle between joint axes z_{i-1} and z_i measured in the direction of axis x_i;

a_i - the common normal distance between joints i and i+1;

d_i - the distance along joint axis z_{i-1} between the origin of system (i-1) and the point of intersection of joint axis z_{i-1} with the common normal between axes z_{i-1} and z_i.

The upper (3×3) submatrix of $^{i-1}T_i$ transforms vectors expressed in system i into vectors expressed in system i-1. The last column of matrix $^{i-1}T_i$ is the position vector between the origins of systems i and i-1, expressed also with respect to system i-1.

If joint i is a sliding one, and the coordinate frame is assigned to the link according to Fig. 1.17 (Subsection 1.3.2), the corresponding homogeneous transformation matrix $^{i-1}T_i$ is

$$^{i-1}T_i = \begin{bmatrix} \cos\theta_i & -\sin\theta_i\cos\alpha_i & \sin\theta_i\sin\alpha_i & 0 \\ \sin\theta_i & \cos\theta_i\cos\alpha_i & -\cos\theta_i\sin\alpha_i & 0 \\ 0 & \sin\alpha_i & \cos\alpha_i & q_i \\ 0 & 0 & 0 & 1 \end{bmatrix} \qquad (2.2.2)$$

where: q_i - is the joint coordinate - the distance between origins of coordinate systems i-1 and i, measured along z_{i-1};

α_i - the twist angle about axis x_i between axes z_{i-1} and z_i;

θ_i - the constant rotation angle about axis z_{i-1} (the previous joint axis) measured between axes x_{i-1} and x_i.

Once the link coordinate systems have been assigned to the links of a given manipulator, parameters α, a, d (for revolute joints) and α and θ for sliding joints, have to be specified. Based on these parameters, the constant values of sines and cosines of angles α and θ are evaluated. The corresponding symbolic forms of the elements of transformation matrices $^{i-1}T_i$, i=1,...,n can be readily obtained, according to Equations (2.2.1) and (2.2.2).

Here, we are interested in obtaining symbolic expressions for the elements of the homogeneous transformation matrix $^{O}T_n$, relating between the coordinate system of link n and the base system. This matrix is obtained as matrix product

$$^{O}T_n = {}^{O}T_1 \, {}^{1}T_2 \, \cdots \, {}^{n-1}T_n \qquad (2.2.3)$$

The matrix multiplication in the above equation can be performed either by starting from the right side and successively premultiplying by matrices $^{i-1}T_i$, $i=n-1,\ldots,1$, or by starting from the left and successively postmultiplying by matrices $^{i-1}T_i$, $i=2,\ldots,n$. The direction of recursiveness is rather important in the process of automatic symbolic model generation, especially in deriving Jacobian elements.

In the case when premultiplying is carried out, i.e. $^{o}T_n$ computed as

$$^{o}T_n = {}^{o}T_1({}^{1}T_2(\ldots({}^{n-2}T_{n-1}{}^{n-1}T_n)))$$ (2.2.4)

the following backward relations are used

$$^{i-1}T_n = {}^{i-1}T_i\,{}^{i}T_n, \qquad i=n-1,\ldots,1$$ (2.2.5)

The upper (3×3) submatrix of $^{i-1}T_n$ describes the rotation of coordinate system n with respect to coordinate system i-1. The fourth column of $^{i-1}T_n$ corresponds to the position vector from the origin of system i-1 to manipulator tip (origin of system n), expressed in frame i-1. Since the origins of link coordinate systems coincide with joint centers, the fourth columns of the obtained matrices $^{i-1}T_n$, $i=n-1,\ldots,1$ are directly applicable in Jacobian matrix generation.

On the other hand, one can obtain matrix $^{o}T_n$ according to

$$^{o}T_n = (((^{o}T_1\,{}^{1}T_2)\ldots)^{n-2}T_{n-1})^{n-1}T_n$$ (2.2.6)

or, equivalently, using the forward recursive relations

$$^{o}T_i = {}^{o}T_{i-1}\,{}^{i-1}T_i, \qquad i=2,\ldots,n$$ (2.2.7)

The upper (3×3) submatrix of $^{o}T_i$ describes the rotation of coordinate system i with respect to the base, reference frame, while the fourth column represents the position vector from the origin of the base system to the end of link i, expressed in the base reference system. The upper (3×3) submatrices of $^{o}T_i$, $i=1,\ldots,n$ are necessary for deriving symbolic expressions for Jacobian matrix ^{o}J with respect to the base system, as will be shown in Section 2.4.

So far as computation of $^{o}T_n$ matrix is considered, the backward and forward recursive relations are practically equal in computational complexity. The only difference appears in the method of grouping variables. However, if the Jacobian together with $^O T_n$ is to be computed, the type of recursiveness plays an important role in reducing, computational complexity. Methods of achieving the minimal number of floating--point operations will be presented in Sections 2.3 and 2.4.

In order to illustrate the differences appearing when applying either backward or forward recursive relations in direct problem solving, we will derive symbolic kinematic equations for a six degree-of-freedom manipulator in the next subsection. These manually derived expressions will be compared later with computer-generated expressions.

2.2.2. Kinematic equations for the UMS-3B manipulator

Figure 2.1 shows the industrial robot GORO 102 (its basic version UMS--3B has been designed in "Mihailo Pupin" Institute, Belgrade, Yugoslavia). In Figure 2.2 the UMS-3B manipulator is shown with coordinate frames assigned to the links, in its initial configuration when all joint coordinate of the revolute joints are equal to zero.

We will use the following abbreviations for the sines and cosines of the angles q_i, α_i and θ_i

$$\sin q_i = S_i, \quad \cos q_i = C_i,$$

$$\sin \alpha_i = SA_i, \quad \cos \alpha_i = CA_i, \qquad i=1,\ldots,n \qquad (2.2.8)$$

$$\sin \theta_i = ST_i, \quad \cos \theta_i = CT_i$$

These abbreviations will be used not only for deriving kinematic equations for the UMS-3B robot, but also in the program realization of the algorithm for computer-aided generation of symbolic models. The abbreviations for the sines and cosines of angles α_i and θ_i are introduced only if their values are not equal 0 or ± 1.

The kinematic parameters for this manipulator are shown in Table T.2.1.

The homogeneous transformations $^{i-1}T_i$, $i=1,\ldots,n$ for this manipulator are as follows:

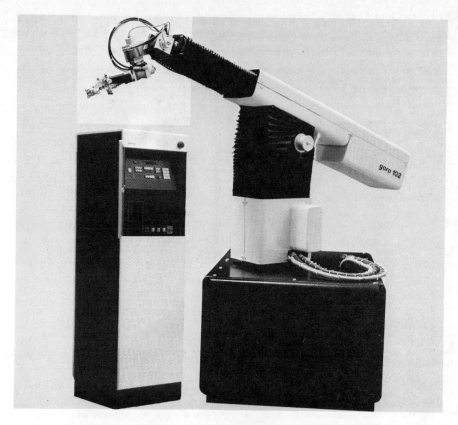

Fig. 2.1. The GORO 102 robot

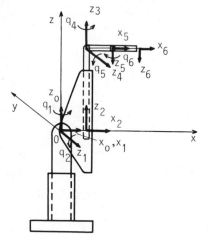

Fig. 2.2. Coordinate frames for
the UMS-3B manipulator

Link	Coordinate	α_i	a_i	d_i	$\cos\alpha_i$	$\sin\alpha_i$	θ_i
1	q_1	90°	0	0	0	1	–
2	q_2	-90°	a_2	0	0	-1	–
3	q_3	0°	0	–	1	0	0°
4	q_4	90°	0	0	0	1	–
5	q_5	90°	a_5	0	0	1	–
6	q_6	0°	a_6	0	1	0	–

T.2.1. Kinematic parameters for the
UMS-3B manipulator

$$^{0}T_1 = \begin{bmatrix} C_1 & 0 & S_1 & 0 \\ S_1 & 0 & -C_1 & 0 \\ 0 & 1 & 0 & 0 \\ 0 & 0 & 0 & 1 \end{bmatrix} \tag{2.2.9}$$

$$^{1}T_2 = \begin{bmatrix} C_2 & 0 & -S_2 & a_2C_2 \\ S_2 & 0 & C_2 & a_2S_2 \\ 0 & -1 & 0 & 0 \\ 0 & 0 & 0 & 1 \end{bmatrix} \tag{2.2.10}$$

$$^{2}T_3 = \begin{bmatrix} 1 & 0 & 0 & 0 \\ 0 & 1 & 0 & 0 \\ 0 & 0 & 1 & q_3 \\ 0 & 0 & 0 & 1 \end{bmatrix} \tag{2.2.11}$$

$$^{3}T_4 = \begin{bmatrix} C_4 & 0 & S_4 & 0 \\ S_4 & 0 & -C_4 & 0 \\ 0 & 1 & 0 & 0 \\ 0 & 0 & 0 & 1 \end{bmatrix} \tag{2.2.12}$$

$$^{4}T_5 = \begin{bmatrix} C_5 & 0 & S_5 & a_5C_5 \\ S_5 & 0 & -C_5 & a_5S_5 \\ 0 & 1 & 0 & 0 \\ 0 & 0 & 0 & 1 \end{bmatrix} \tag{2.2.13}$$

$$^{5}T_6 = \begin{bmatrix} C_6 & -S_6 & 0 & a_6C_6 \\ S_6 & C_6 & 0 & a_6S_6 \\ 0 & 0 & 1 & 0 \\ 0 & 0 & 0 & 1 \end{bmatrix} \tag{2.2.14}$$

We will first consider the application of the backward recursive relation (2.2.5) in solving the direct kinematic problem. Thus we obtain

$$
{}^{5}T_{6} = \begin{bmatrix} c_6 & -s_6 & 0 & a_6c_6 \\ s_6 & c_6 & 0 & a_6s_6 \\ 0 & 0 & 1 & 0 \\ 0 & 0 & 0 & 1 \end{bmatrix}
\tag{2.2.15}
$$

$$
{}^{4}T_{6} = \begin{bmatrix} c_5c_6 & -c_5c_6 & s_5 & c_5(a_6c_6+a_5) \\ s_5c_6 & -s_5s_6 & -c_5 & s_5(a_6c_6+a_5) \\ s_6 & c_6 & 0 & a_6s_6 \\ 0 & 0 & 0 & 1 \end{bmatrix}
\tag{2.2.16}
$$

$$
{}^{3}T_{6} = \begin{bmatrix} c_4c_5c_6+s_4s_6 & -c_4c_5s_6+s_4c_6 & c_4s_5 & c_4c_5(a_6c_6+a_5)+s_4a_6s_6 \\ s_4c_5c_6-c_4s_6 & -s_4c_5s_6-c_4c_6 & s_4s_5 & s_4c_5(a_6c_6+a_5)-c_4a_6s_6 \\ s_5c_6 & -s_5s_6 & -c_5 & s_5(a_6c_6+a_5) \\ 0 & 0 & 0 & 1 \end{bmatrix}
\tag{2.2.17}
$$

$$
{}^{2}T_{6} = \begin{bmatrix} c_4c_5c_6+s_4s_6 & -c_4c_5s_6+s_4c_6 & c_4s_5 & c_4c_5(a_6c_6+a_3)+s_4a_6s_6 \\ s_4c_5c_6-c_4s_6 & -s_4c_5s_6-c_4c_6 & s_4s_5 & s_4c_5(a_6c_6+a_5)-c_4a_6s_6 \\ s_5c_6 & -s_5s_6 & -c_5 & s_5(a_6c_6+a_5)+q_3 \\ 0 & 0 & 0 & 1 \end{bmatrix}
\tag{2.2.18}
$$

$$
{}^{1}T_{6} = \begin{bmatrix} c_2(c_4c_5c_6+s_4s_6)-s_2s_5c_6 & c_2(-c_4c_5s_6+s_4c_6)+s_2s_5s_6 \\ s_2(c_4c_5c_6+s_4s_6)+c_2s_5c_6 & s_2(-c_4c_5s_6+s_4c_6)-c_2s_5s_6 \\ -s_4c_5c_6+c_4s_6 & s_4c_5s_6+c_4c_6 \\ 0 & 0 \end{bmatrix}
$$

$$
\begin{matrix} c_2c_4s_5+s_2c_5 & c_2[c_4c_5(a_6c_6+a_5)+s_4a_6s_6+a_2]-s_2[s_5(a_6c_6+a_5)+q_3] \\ s_2c_4s_5-c_2c_5 & s_2[c_4c_5(a_6c_6+a_5)+s_4a_6s_6+a_2]+c_2[s_5(a_6c_6+a_5)+q_3] \\ -s_4s_5 & -s_4c_5(a_6c_6+a_5)+c_4a_6s_6 \\ 0 & 1 \end{matrix}
\tag{2.2.19}
$$

$$
{}^{0}T_{6} = \begin{bmatrix} T6011 & T6012 & T6013 & T6014 \\ T6021 & T6022 & T6023 & T6024 \\ T6031 & T6032 & T6033 & T6034 \\ 0 & 0 & 0 & 1 \end{bmatrix} \qquad (2.2.20)
$$

where

$$T6011 = C_1[C_2(C_4C_5C_6+S_4S_6)-S_2S_5C_6]-S_1(S_4C_5C_6-C_4S_6)$$

$$T6021 = S_1[C_2(C_4C_5C_6+S_4S_6)-S_2S_5C_6]+C_1(S_4C_5C_6-C_4S_6)$$

$$T6031 = S_2(C_4C_5C_6+S_4S_6)+C_2S_5C_6$$

$$T6012 = C_1[C_2(-C_4C_5S_6+S_4C_6)+S_2S_5S_6]-S_1(-S_4C_5S_6-C_4C_6)$$

$$T6022 = S_1[C_2(-C_4C_5S_6+S_4C_6)+S_2S_5S_6]+C_1(-S_4C_5S_6-C_4C_6)$$

$$T6032 = S_2(-C_4C_5S_6+S_4C_6)-C_2S_5S_6$$

$$T6013 = C_1(C_2C_4S_5+S_2C_5)-S_1S_4S_5 \qquad (2.2.21)$$

$$T6021 = S_1(C_2C_4S_5+S_2C_5)+C_1S_4S_5$$

$$T6033 = S_2C_4S_5-C_2C_5$$

$$T6014 = C_1\{C_2[C_4C_5(a_6C_6+a_5)+S_4a_6S_6+a_2]-S_2[S_5(a_6C_6+a_5)+q_3]\}-$$
$$-S_1[S_4C_5(a_6C_6+a_5)-C_4a_6S_6]$$

$$T6024 = S_1\{C_2[C_4C_5(a_6C_6+a_5)+S_4a_6S_6+a_2]-S_2[S_5(a_6C_6+a_5)+q_3]\}+$$
$$+ C_1[S_4C_5(a_6C_6+a_5)-C_4a_6S_6]$$

$$T6034 = S_2[C_4C_5(a_6C_6+a_5)+S_4a_6S_6+a_2]+C_2[S_5(a_6C_6+a_5)+q_3]$$

It may be seen from the above equations that the elements of matrix ${}^{0}T_{6}$ are relatively complex. In order to compute the whole matrix, we require 10 transcendental function calls, 54 multiplies, and 25 additions. The left three columns take 38 multiplies and 16 additions, while the position vector takes 16 multiplies and 9 additions. One of three first columns could also be obtained as the vector cross product of the remaining two columns, due to the orthogonality of the rotation matrix. This would reduce the number of floating-point multiplications and additions, but only as far as the evaluation of $^{O}T_6$ is considered. If the evaluation of the Jacobian follows the computation of matrix $^{O}T_6$, it is better not to compute one of the columns numerically. This is due to the fact that the terms which would not be evaluated in that case, and which correspond to matrices $^{1}T_6$, $^{2}T_6$, etc, are necessary for the Jacobian matrix evaluation.

It is evident from expressions (2.2.21) that the variables are grouped starting from the last link, which is due to the application of back-ward recursive relations.

The expressions, equivalent to (2.2.21), can be obtained by applying forward recursive relations (2.2.7), too

$$
^{O}T_1 =
\begin{bmatrix}
C_1 & 0 & S_1 & 0 \\
S_1 & 0 & -C_1 & 0 \\
0 & 1 & 0 & 0 \\
0 & 0 & 0 & 1
\end{bmatrix}
\tag{2.2.22}
$$

$$
^{O}T_2 =
\begin{bmatrix}
C_2C_1 & -S_1 & -S_2C_1 & a_2C_2C_1 \\
C_2S_1 & C_1 & -S_2S_1 & a_2C_2S_1 \\
S_2 & 0 & C_2 & a_2S_2 \\
0 & 0 & 0 & 1
\end{bmatrix}
\tag{2.2.23}
$$

$$
^{O}T_3 =
\begin{bmatrix}
C_2C_1 & -S_1 & -S_2C_1 & -q_3S_2C_1+a_2C_2C_1 \\
C_2S_1 & C_1 & -S_2S_1 & -q_3S_2S_1+a_2C_2S_1 \\
S_2 & 0 & C_2 & q_3C_2+a_2S_2 \\
0 & 0 & 0 & 1
\end{bmatrix}
\tag{2.2.24}
$$

$$
^{O}T_4 =
\begin{bmatrix}
C_4C_2C_1-S_4S_1 & -S_2C_1 & S_4C_2C_1+C_4S_1 & -q_3S_2C_1+a_2C_2C_1 \\
C_4C_2S_1+S_4C_1 & -S_2S_1 & S_4C_2S_1-C_4C_1 & -q_3S_2S_1+a_2C_2S_1 \\
C_4S_2 & C_2 & S_4S_2 & q_3C_2+a_2S_2 \\
0 & 0 & 0 & 1
\end{bmatrix}
\tag{2.2.25}
$$

$$
{}^{o}T_5 = \begin{bmatrix}
C_5(C_4C_2C_1-S_4S_1)-S_5S_2C_1 & S_4C_2C_1+C_4S_1 & S_5(C_4C_2C_1-S_4S_1)+C_5S_2C_1 \\
C_5(C_4C_2S_1+S_4C_1)-S_5S_2S_1 & S_4C_2S_1-C_4C_1 & S_5(C_4C_2S_1+S_4C_1)+C_5S_2S_1 \\
C_5C_4S_2+S_5C_2 & S_4S_2 & S_5C_4S_2-C_5C_2 \\
0 & 0 & 0
\end{bmatrix}
$$

$$
\left.\begin{array}{c}
a_5[C_5(C_4C_2C_1-S_4S_1)-S_5S_2C_1]-q_3S_2C_1+a_2C_2C_1 \\
a_5[C_5(C_4C_2S_1+S_4C_1)-S_5S_2S_1]-q_3S_2S_1+a_2C_2S_1 \\
a_5(C_5C_4S_2+S_5C_2)+q_3C_2+a_2S_2 \\
1
\end{array}\right] \qquad (2.2.26)
$$

$$
{}^{o}T_6 = \begin{bmatrix}
T6011 & T6012 & T6013 & T6014 \\
T6021 & T6022 & T6023 & T6024 \\
T6031 & T6032 & T6033 & T6034 \\
0 & 0 & 0 & 1
\end{bmatrix} \qquad (2.2.27)
$$

$T6011 = C_6[C_5(C_4C_2C_1-S_4S_1)-S_5S_2C_1]+S_6(S_4C_2C_1+C_4S_1)$

$T6021 = C_6[C_5(C_4C_2S_1+S_4C_1)-S_5S_2S_1]+S_6(S_4C_2S_1-C_4C_1)$

$T6031 = C_6(C_5C_4S_2+S_5C_2)+S_6S_4S_2$

$T6012 = -S_6[C_5(C_4C_2C_1-S_4S_1)-S_5S_2C_1]+C_6(S_4C_2C_1+C_4S_1)$

$T6022 = -S_6[C_5(C_4C_2S_1+S_4C_1)-S_5S_2S_1]+C_6(S_4C_2S_1-C_4C_1)$

$T6032 = -S_6(C_5C_4S_2+S_5C_2)+C_6S_4S_2$ $\qquad (2.2.28)$

$T6013 = S_5(C_4C_2C_1-S_4S_1)+C_5S_2C_1$

$T6023 = S_5(C_4C_2S_1+S_4C_1)+C_5S_2S_1$

$T6033 = S_5C_4S_2-C_5C_2$

$T6014 = a_6\{C_6[C_5(C_4C_2C_1-S_4S_1)-S_5S_2C_1]+S_6(S_4C_2C_1+C_4S_1)\}+$

$\qquad + a_5[C_5(C_4C_2C_1-S_4S_1)-S_5S_2C_1]-q_3S_2C_1+a_2C_2C_1$

$T6024 = a_6\{C_6[C_5(C_4C_2S_1+S_4C_1)-S_5S_2S_1]+S_6(S_4C_2S_1-C_4C_1)\}+$

$\qquad + a_5[C_5(C_4C_2S_1+S_4C_1)-S_5S_2S_1]-q_3S_2S_1+a_2C_2S_1$

$$T6034 = a_6[C_6(C_5C_4S_2+S_5C_2)+S_6S_4S_2]+a_5(C_5C_4S_2+S_5C_2)+q_3C_2+a_2S_2$$

The evaluation of matrix $^0T_6$ according to these equations requires 38 multiplications and 16 additions for the left three columns, and 12 multiplies and 9 additions for the position vector. We can see that the number of operations necessary for the position vector is less than in the case of equations (2.2.21). However, one should note that the fourth columns of matrices ^0T_i i=1,...,n-1, as opposed to the fourth columns of matrices $^{i-1}T_n$, i=n,...,1, are not required in the evaluation of the Jacobian which corresponds to the manipulator tip. This fact will be taken into account in computer-aided generation of the Jacobian with respect to the base reference system (Section 2.4).

Equations (2.2.21) and (2.2.28) show that their derivation is a boring and tedious task, subject to human error. Besides, once the expressions have been derived, the corresponding optimal procedure for their calculation should be written down. Therefore, it is very convenient to obtain the solution automatically, with the aid of a computer.

2.2.3. Backward recursive symbolic relations

In this subsection we will derive backward recursive relations which yield symbolic expressions for the elements of homogeneous transformation matrices $^{i-1}T_n$, i=1,...,n. These relations are directly utilized in the program implementation of the software package for symbolic kinematic model generation.

In this chapter we will introduce the following notation for the elements of the transformation matrices ^iT_j

$$^iT_j = \begin{bmatrix} Tji11 & Tji12 & Tji13 & Tji14 \\ Tji21 & Tji22 & Tji23 & Tji24 \\ Tji31 & Tji32 & Tji33 & Tji34 \\ 0 & 0 & 0 & 1 \end{bmatrix} \qquad (2.2.29)$$

This notation is suitable from the standpoint of program implementation of the algorithm.

We will start from the general backward recursive relation (2.2.5), graphically outlined in Fig. 2.3.

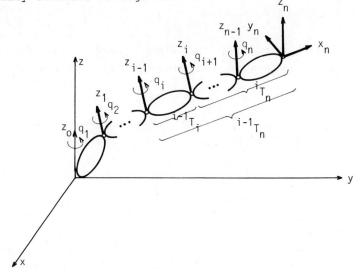

Fig. 2.3. An illustration of the backward recursive relations

Let us first consider the case when joint i is a revolute one. If matrix ^iT_n has already been evaluated, the matrix $^{i-1}T_n$ is, according to (2.2.1) and (2.2.5), given by

$$^{i-1}T_n = \begin{bmatrix} \cos q_i & -\sin q_i \cos\alpha_i & \sin q_i \sin\alpha_i & a_i \cos q_i \\ \sin q_i & \cos q_i \cos\alpha_i & -\cos q_i \sin\alpha_i & a_i \sin q_i \\ 0 & \sin\alpha_i & \cos\alpha_i & d_i \\ 0 & 0 & 0 & 1 \end{bmatrix} {}^iT_n \quad (2.2.30)$$
$$1<i<n$$

For i=n, the matrix ^nT_n is assummed to be the (4×4) unit matrix. If we introduce the notation (2.2.29) for the elements of matrix ^iT_n and $^{i-1}T_n$, we can write the following recursive relations for the elements of matrix $^{i-1}T_n$, $1<i<n$

$$Tn(i-1)1\ell = \cos q_i Tni1\ell - \sin q_i [\cos\alpha_i Tni2\ell - \sin\alpha_i Tni3\ell], \quad \ell=1,2,3 \quad (2.2.31)$$

$$Tn(i-1)14 = \cos q_i (Tni14+a_i) - \sin q_i [\cos\alpha_i Tni24 - \sin\alpha_i Tni34] \quad (2.2.32)$$

$$Tn(i-1)2\ell = \sin q_i Tni1\ell + \cos q_i [\cos\alpha_i Tni2\ell - \sin\alpha_i Tni3\ell], \quad \ell=1,2,3 \quad (2.2.33)$$

$$Tn(i-1)24 = \sin q_i (Tni14+a_i) + \cos q_i [\cos\alpha_i Tni24 - \sin\alpha_i Tni34] \quad (2.2.34)$$

$$Tn(i-1)3\ell = \sin\alpha_i Tni2\ell + \cos\alpha_i Tni3\ell, \qquad \ell=1,2,3 \qquad\qquad (2.2.35)$$

$$Tn(i-1)34 = \sin\alpha_i Tni24 + \cos\alpha_i Tni34 + d_i \qquad\qquad (2.2.36)$$

As we can see, these are common terms which appear in (2.2.31)-(2.2.34). These terms should be evaluated in advance in order to reduce the number of multiplications and additions. Therefore, we introduce new scalar variables

$$TRni\ell = \cos\alpha_i Tni2\ell - \sin\alpha_i Tni3\ell, \qquad \ell=1,2,3,4, \quad 1<i<n \qquad (2.2.37)$$

and

$$TPni = Tni14 + a_i, \qquad 1<i<n \qquad\qquad (2.2.38)$$

In that case recursive symbolic relations (2.2.31)-(2.2.36) simplify and become

$$Tn(i-1)1\ell = \cos q_i Tni1\ell - \sin q_i TRni\ell, \qquad \ell=1,2,3 \qquad\qquad (2.2.39)$$

$$Tn(i-1)14 = \cos q_i TPni - \sin q_i TRni4 \qquad\qquad (2.2.40)$$

$$Tn(i-1)2\ell = \sin q Tni1\ell + \cos q_i TRni\ell, \qquad \ell=1,2,3 \qquad\qquad (2.2.41)$$

$$Tn(i-1)24 = \sin q_i TPni + \cos q_i TRni4 \qquad\qquad (2.2.42)$$

$$Tn(i-1)3\ell = \sin\alpha_i Tni2\ell + \cos\alpha_i Tni3\ell, \qquad \ell=1,2,3 \qquad\qquad (2.2.43)$$

$$Tn(i-1)34 = \sin\alpha_i Tni24 + \cos\alpha_i Tni34 + d_i \qquad 1<i<n \qquad (2.2.44)$$

We will illustrate the use of these relations later in this subsection.

Let us now consider the case when joint i is prismatic. The homogeneous transformation matrix $^{i-1}T_i$, corresponding to the joint, has the form (2.2.2), where θ_i and α_i are constant parameters. We will assume that the constant elements of $^{i-1}T_i$ have been evaluated and stored into variables Ti(i-1)jℓ, j=1,2,3, ℓ=1,2,3. In that case the backward recursive relation (2.2.5) becomes

$$
{}^{i-1}T_n =
\begin{bmatrix}
Ti(i-1)11 & Ti(i-1)12 & Ti(i-1)13 & 0 \\
Ti(i-1)21 & Ti(i-1)22 & Ti(i-1)23 & 0 \\
Ti(i-1)31 & Ti(i-1)32 & Ti(i-1)33 & q_i \\
0 & 0 & 0 & 1
\end{bmatrix}
{}^{i}T_n, \quad 1<i<n
$$

(2.2.45)

yielding the following recursive relations for the elements of ${}^{i-1}T_n$, $1<i<n$ (here * stands for multiplication)

Tn(i-1)jℓ = Ti(i-1)j1*Tni1ℓ + Ti(i-1)j2*Tni2ℓ +

+ Ti(i-1)j3*Tni3ℓ,   j=1,2,3,   ℓ=1,2,3   (2.2.46)

Tn(i-1)j4 = Ti(i-1)j1*Tni14 + Ti(i-1)j2*Tni24 +

+ Ti(i-1)j3*Tni34,   j=1,2   (2.2.47)

Tn(i-1)34 = Ti(i-1)32*Tni24 + Ti(i-1)33*Tni34 + q_i (2.2.48)

With industrial manipulators it often happens that some elements Ti(i-1)jℓ are either zero or ±1 for a matrix ${}^{i-1}T_i$ which corresponds to a prismatic joint. This will obviously be taken into account in the program implementation of the algorithm in order to avoid needless multiplications and additions. The same examination on 0 or ±1 is also carried out on all the other variables figuring in the equations.

By applying either the relations (2.2.39) - (2.2.44), or relations (2.2.46) - (2.2.48), successively for i=n, n-1,...,1, we can obtain a set of equations to evaluate the transformation matrix ${}^{O}T_n$ optimally with respect to the number of multiplications and additions.

Let us now illustrate the application of the derived symbolic recursive relations by solving the direct kinematic problem for the UMS-3B manipulator. The kinematic parameters are shown in Table T.2.1.

In the following text we will use the term - *complex variable*. The evaluation of the complex variables requires at least one additional multiplication or at least one additional addition. On the other hand, a *simple variable* is either equal 0, ±1 or is equal to another variable or parameter whose value has already been obtained. The computation of a simple variable requires no additional floating-point operation.

For the UMS-3B manipulator n equals 6 and the last joint is a revolute one with $\cos\alpha_6 = 1$, $\sin\alpha_6 = 0$. Applying equations (2.2.37) - (2.2.44) for i=n and knowing that ${}^6T_6 = I_4$, we can write

$$TR661 = 0$$

$$TR662 = 1$$

$$TR663 = 0$$

$$TR664 = 0$$

$$TP66 = a_6$$

$$T6511 = C_6$$

$$T6521 = S_6$$

$$T6531 = 0$$

$$T6512 = -S_6$$

$$T6522 = C_6$$

$$T6532 = 0$$

$$T6513 = 0$$

$$T6523 = 0$$

$$T6533 = 1$$

$$T6514 = C_6 * a_6$$

$$T6524 = S_6 * a_6$$

$$T6534 = 0$$

(2.2.49)

Thus we have obtained the elements of matrix ${}^5T_6$ which are equivalent to (2.2.14). The only complex elements are T6514 and T6524. The sign * stands for multiplication. The abbreviations (2.2.8) have been used.

Joint 5 is also revolute, with $\cos\alpha_5 = 0$ and $\sin\alpha_5 = 1$. By applying relations (2.2.37) - (2.2.44) for i=5 we obtain the elements of matrix ${}^4T_6$

$$TR651 = 0$$

$$TR652 = 0$$

$$TR653 = -1$$

$$TR654 = 0$$

$$TP65 = T6514 + a_5$$

$$T6411 = C_5 * C_6$$

$$T6421 = S_5*C_6$$

$$T6431 = S_6$$

$$T6412 = -C_5*S_6$$

$$T6422 = -S_5*S_6$$

$$T6432 = C_6$$

$$T6413 = S_5$$

$$T6423 = -C_5$$

$$T6433 = 0$$

$$T6414 = C_5*TP65$$

$$T6424 = S_5*TP65$$

$$T6434 = T6524$$

(2.2.50)

We may note that these equations were written using the following principle: the simple variables on the right hand sides of the equations are replaced by their real contents. For example, instead of writing

$$T6411 = C_5*T6511 \qquad (2.2.51)$$

we wrote $T6411 = C_5*C_6$, since the real content of the variable $T6511$ is C_6 (see Equations (2.2.49)).

The set of equations for evaluating the elements of matrix $^3T_6$ are also obtained from relations (2.2.37) - (2.2.44) for $i=4$

$$TR641 = -S_6$$

$$TR642 = -C_6$$

$$TR643 = 0$$

$$TR644 = -T6524$$

$$TR64 = T6414$$

$$T6311 = C_4*T6411+S_4*S_6$$

$$T6321 = S_4*T6411-C_4*S_6$$

$$T6331 = T6421$$

$$T6312 = C_4*T6412+S_4*C_6$$

$$T6322 = S_4*T6412-C_4*C_6$$

$$T6323 = T6421$$

(2.2.52)

$$T6313 = C_4 * S_5$$

$$T6323 = S_4 * S_5$$

$$T6333 = -C_5$$

$$T6314 = C_4 * T6414 + S_4 * T6524$$

$$T6324 = S_4 * T6414 - C_4 * T6524$$

$$T6334 = T6424$$

The homogeneous transformation matrix $^2T_6$ is obtained from relations (2.2.46)-(2.2.48) for i=3 since the third degree of freedom is a sliding one

$$T6211 = T6311$$

$$T6221 = T6321$$

$$T6231 = T6331$$

$$T6212 = T6312$$

$$T6222 = T6322$$

$$T6232 = T6422 \qquad\qquad (2.2.53)$$

$$T6213 = T6313$$

$$T6223 = T6323$$

$$T6233 = -C_5$$

$$T6214 = T6314$$

$$T6224 = T6324$$

$$T6234 = T6424 + q_3$$

The above mentioned principle of replacing the simple variables by their real contents also makes it possible to avoid multiplying by the variables which are, in fact, always equal to zero or ±1 and to avoid adding zero variables.

The elements of matrix $^1T_6$ are obtained by further application of the symbolic relations (2.2.37) - (2.2.44) for i=2

$$TR621 = T6421$$

$$TR622 = T6422$$

$$TR623 = -C_5$$

TR624 = T6234

TP62 = T6314+a_2

T6111 = C_2*T6311-S_2*T6421

T6121 = S_2*T6311+C_2*T6421

T6131 = -T6321

T6112 = C_2*T6312-S_2*T6422

T6122 = S_2*T6312+C_2*T6422   (2.2.54)

T6132 = -T6322

T6113 = C_2*T6313+S_2*C_5

T6123 = S_2*T6313-S_2*C_5

T6133 = -T6323

T6114 = C_2*TP62-S_2*T6234

T6124 = S_2*TP62+C_2*T6234

T6134 = -T6324

Finally, we obtain the equations which will yield the elements of the homogeneous transformation between the end-effector coordinate frame and the base frame

TR611 = T6321

TR612 = T6322

TR613 = T6323

TR614 = T6114

TP61 = T6114

T6011 = C_1*T6111-S_1*T6321

T6021 = S_1*T6111+C_1*T6321

T6031 = T6121   (2.2.55)

T6012 = C_1*T6112-S_1*T6322

T6022 = S_1*T6112+C_1*T6322

T6032 = T6122

T6013 = C_1*T6113-S_1*T6323

T6023 = S_1*T6113+C_1*T6323

T6033 = T6123

$$T6014 = C_1*T6114-S_1*T6324$$

$$T6024 = S_1*T6114+C_1*T6324$$

$$T6034 = T6124$$

The set of equations (2.2.49) - (2.2.55) yield the direct kinematic problem solution for the UMS-3B manipulator. If these equations were written by a computer, the lines which contain no multiplications or additions would be omitted, except for those lines which correspond to the elements of matrix ${}^OT_6$.

We may observe that the recursive symbolic relations yielding the direct kinematic problem solution allow for no compression of the trigonometric expressions. This is due to the fact that there are no multiplications (or additions) between matrices (or vectors) which are the functions of the same joint coordinates. This will not be the case with the Jacobian matrix computation.

2.2.4. Forward recursive symbolic relations

In this subsection we will derive forward recursive relations which yield symbolic expressions for the elements of transformation matrices ${}^OT_i$, i=1,...,n. They ensure optimal evaluation of the elements of these matrices.

We will derive these symbolic relations starting from the general forward recursive relation (2.2.7), outlined in Fig. 2.4.

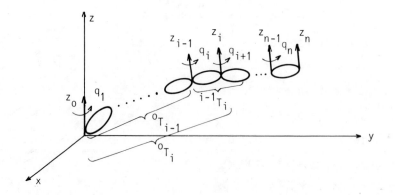

Fig. 2.4. A scheme of forward recursive relations

Let us assume that the matrix $^{O}T_{i-1}$ has already been evaluated, and that joint i is revolute. According to equations (2.2.7) and (2.2.1) we can write

$$
^{O}T_i = {^{O}T_{i-1}} \begin{bmatrix} \cos q_i & -\sin q_i \cos \alpha_i & \sin q_i \sin \alpha_i & a_i \cos q_i \\ \sin q_i & \cos q_i \cos \alpha_i & -\cos q_i \sin \alpha_i & a_i \sin q_i \\ 0 & \sin \alpha_i & \cos \alpha_i & d_i \\ 0 & 0 & 0 & 1 \end{bmatrix}
\qquad (2.2.56)
$$

$$1 \leqslant i \leqslant n$$

For $i=1$, the matrix $^{O}T_{O}$ is assumed to be a (4×4) unit matrix. If we introduce the notation (2.2.29) for the elements of $^{O}T_{i-1}$ and $^{O}T_i$, we can write the following recursive relations for the elements of matrix $^{O}T_i$, $1 \leqslant i \leqslant n$

$$Tioj1 = \cos q_i T(i-1)oj1 + \sin q_i T(i-1)oj2, \qquad j=1,2,3 \qquad (2.2.57)$$

$$Tioj2 = -\cos \alpha_i (\sin q_i T(i-1)oj1 - \cos q_i T(i-1)oj2) +$$
$$+ \sin \alpha_i T(i-1)oj3, \qquad\qquad j=1,2,3 \qquad (2.2.58)$$

$$Tioj3 = \sin \alpha_i (\sin q_i T(i-1)oj1 - \cos q_i T(i-1)oj2) +$$
$$+ \cos \alpha_i T(i-1)oj3, \qquad\qquad j=1,2,3 \qquad (2.2.59)$$

$$Tioj4 = a_i (\cos q_i T(i-1)oj1 + \sin q_i T(i-1)oj2) +$$
$$+ d_i T(i-1)oj3 + T(i-1)oj4, \qquad\qquad j=1,2,3 \qquad (2.2.60)$$

The term common to equations (2.2.58) and (2.2.59) should be evaluated in advance in order to reduce the number of operations. Thus, we introduce new scalar variables

$$TS(i-1)oj = \sin q_i T(i-1)oj1 - \cos q_i T(i-1)oj2, \qquad j=1,2,3$$
$$1 \leqslant i \leqslant n \qquad (2.2.61)$$

Besides, the variables Tioj4 can be evaluated using the already evaluated elements Tioj1. Accordingly, the equations (2.2.57) - (2.2.60) become

$$Tioj1 = \cos q_i T(i-1)oj1 + \sin q_i T(i-1)oj2, \qquad j=1,2,3 \qquad (2.2.62)$$

$$\text{Tioj2} = -\cos\alpha_i \text{TS}(i-1)\text{oj} + \sin\alpha_i \text{T}(i-1)\text{oj3}, \qquad j=1,2,3 \qquad (2.2.63)$$

$$\text{Tioj3} = \sin\alpha_i \text{TS}(i-1)\text{oj} + \cos\alpha_i \text{T}(i-1)\text{oj3}, \qquad j=1,2,3 \qquad (2.2.64)$$

$$\text{Tioj4} = a_i \text{Tioj1} + d_i \text{T}(i-1)\text{oj3} + \text{T}(i-1)\text{oj4}, \qquad j=1,2,3$$

$$1 \leqslant i \leqslant n \qquad (2.2.65)$$

These expressions represent the forward recursive symbolic relations enabling optimal computation of matrix $^{O}T_i$, if joint i is revolute and if matrix $^{O}T_{i-1}$ has already been evaluated.

Let us now consider the computation of $^{O}T_i$ if joint i is prismatic. We will here assume, as before, that the constant elements of $^{i-1}T_i$ have been calculated and stored into variables Ti(i-1)jℓ, j=1,2,3, ℓ=1,2,3. Then, the general reccurent relation (2.2.7) becomes

$$^{O}T_i = {}^{O}T_{i-1} \begin{bmatrix} \text{Ti}(i-1)11 & \text{Ti}(i-1)12 & \text{Ti}(i-1)13 & 0 \\ \text{Ti}(i-1)21 & \text{Ti}(i-1)22 & \text{Ti}(i-1)23 & 0 \\ \text{Ti}(i-1)31 & \text{Ti}(i-1)32 & \text{Ti}(i-1)33 & q_i \\ 0 & 0 & 0 & 1 \end{bmatrix} \quad 1 \leqslant i \leqslant n \quad (2.2.66)$$

yielding the following recursive relations for the elements of $^{O}T_i$, $1 \leqslant i \leqslant n$

$$\text{Tiojℓ} = \text{T}(i-1)\text{oj1} * \text{Ti}(i-1)1ℓ + \text{T}(i-1)\text{oj2} * \text{Ti}(i-1)2ℓ +$$

$$+ \text{T}(i-1)\text{oj3} * \text{Ti}(i-1)3ℓ, \qquad j=1,2,3, \qquad ℓ=1,2,3 \qquad (2.2.67)$$

$$\text{Tioj4} = \text{T}(i-1)\text{oj3} * q_i + \text{T}(i-1)\text{oj4}, \quad j=1,2,3, \qquad 1 \leqslant i \leqslant n \qquad (2.2.68)$$

These expressions, together with (2.2.61)-(2.2.65), represent the basis for automatic, computer-aided generation of computer programs for evaluating transformation matrices $^{O}T_i$, i=1,...,n.

We should note, however, that if the computation of $^{O}T_n$ is followed by the computation of the Jacobian ^{O}J with respect to manipulator base, then it is computationally more efficient to evaluate the first three columns of $^{O}T_i$, i=1,...,n according to Equations (2.2.62) – (2.2.64), (2.2.67), while the position vectors should be calculated by backward recursive relations (2.2.40), (2.2.42), (2.2.44) and (2.2.47), (2.2.48). More detailed discussion on this problem will be given in Section 2.4.

We will not illustrate the application of these relations, since it is analogous to the application of the backward recursive relations, presented in the preceding subsection. Besides, this example will be shown in Subsection 2.6.2 in the form of a computer program.

2.2.5. Treatment of the revolute joints with parallel joint axes

In this subsection we will deal with revolute joints with parallel joint axes, within the scope of the direct kinematic problem. Such degrees of freedom appear frequently in contemporary industrial robots.

The direct problem could be solved for such manipulators by applying the general recursive symbolic relations, derived in the preceding subsections, i.e. as if the axes' parallelism has been ignored. However, such an approach would not ensure the minimal number of numeric operations, upon which we insist in this analysis. Therefore, we will develop special recursive symbolic relations for the elements of transformation matrices which include joints with parallel axes. Both backward and forward recursive relations will be derived.

Backward recursive relations

Let us first consider the direct kinematic problem solution according to backward resursive relations. Denote by k the number of the joint whose axis (axis z_{k-1}) is the first in the series of parallel joint axes (see Fig. 2.5). Denote by K the number of the joint whose axis (axis z_{K-1}) is last in the series of parallel axes. The inequality $1 \leqslant k < K \leqslant n$ obviously holds.

We will assume that the directions of all parallel joint axes coincide, i.e. that the axes coincide, i.e. that the axes are chosen in such a way that angles $\alpha_i = 0$, $i = k, k+1, \ldots, K-1$ (but cannot take values $\pm 180^\circ$). This assumption simplifies the analysis, but does not reduce the generality of the expressions derived. In that case, the transformation matrices depend only on the sum of joint angles (and not on the difference). We will use the following abbreviations in the text to follow

$$S_{i,K} = \sin(q_i + q_{i+1} + \cdots + q_{K-1} + q_K)$$

$$c_{i,K} = \cos(q_i + q_{i+1} + \cdots + q_{K-1} + q_K), \qquad 1 \leqslant k \leqslant i < K \leqslant n \qquad (2.2.69)$$

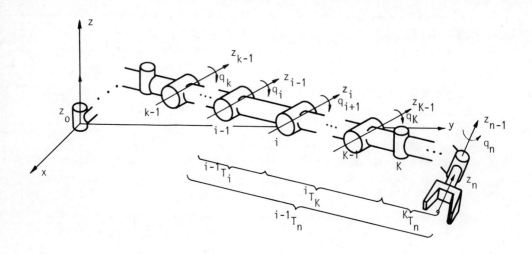

Fig. 2.5. An illustration of the backward recursive relations for joints with parallel axes

By applying the backward recursive relations derived in Subsection 2.2.3, successively for $i=n$, $n-1,\ldots,K$ we can evaluate the elements of matrices ${}^{n-1}T_n$, ${}^{n-2}T_n,\ldots,$ ${}^{K}T_n$, ${}^{K-1}T_n$. In order to obtain matrix ${}^{K-2}T_n$ we will not, as usually, use ${}^{K-2}T_n = {}^{K-2}T_{K-1}\,{}^{K-1}T_n$. We will rather first obtain the product

$$^{K-2}T_K = {}^{K-2}T_{K-1}\,{}^{K-1}T_K \qquad (2.2.70)$$

and then postmultiply by ${}^{K}T_n$ to obtain ${}^{K-2}T_n$

$$^{K-2}T_n = {}^{K-2}T_K\,{}^{K}T_n \qquad (2.2.71)$$

Bearing in mind that axes z_{K-1} and z_{K-2} are parallel (i.e. $\alpha_{K-1}=0$), while α_K has an arbitrary value, the equation (2.2.70) can be written in the following form

$$^{K-2}T_K = \begin{bmatrix} \cos q_{K-1} & -\sin q_K & 0 & a_{K-1}\cos q_{K-1} \\ \sin q_{K-1} & \cos q_K & 0 & a_{K-1}\sin q_{K-1} \\ 0 & 0 & 1 & d_{K-1} \\ 0 & 0 & 0 & 1 \end{bmatrix} \times$$

$$x \begin{bmatrix} \cos q_K & -\sin q_K \cos \alpha_K & \sin q_K \sin \alpha_K & a_K \cos q_K \\ \sin q_K & \cos q_K \cos \alpha_K & \cos q_K \sin \alpha_K & a_K \sin q_K \\ 0 & \sin \alpha_K & \cos \alpha_K & d_K \\ 0 & 0 & 0 & 1 \end{bmatrix} \qquad (2.2.72)$$

or, according to notation (2.2.69), we have

$$^{K-2}T_K = \begin{bmatrix} C_{K-1,K} & -S_{K-1,K}\cos\alpha_K & S_{K-1,K}\sin\alpha_K & a_K C_{K-1,K}+a_{K-1}C_{K-1} \\ S_{K-1,K} & C_{K-1,K}\cos\alpha_K & C_{K-1,K}\sin\alpha_K & a_K S_{K-1,K}+a_{K-1}S_{K-1} \\ 0 & \sin\alpha_K & \cos\alpha_K & d_K+d_{K-1} \\ 0 & 0 & 0 & 1 \end{bmatrix}$$

$$(2.2.73)$$

If there are more than two revolute joints with parallel joint axes, the Equations (2.2.70)-(2.2.73) can be generalized and become

$$^{i-1}T_K = {}^{i-1}T_i \, {}^iT_K, \qquad 1 \leqslant k \leqslant i \leqslant K-1 \leqslant n \qquad (2.2.74)$$

$$^{i-1}T_n = {}^{i-1}T_K \, {}^KT_n \qquad (2.2.75)$$

where

$$^{i-1}T_K = \begin{bmatrix} C_{i,K} & -S_{i,K}\cos\alpha_K & S_{i,K}\sin\alpha_K & a_K C_{i,K}+a_{K-1}C_{i,K-1}+\ldots+a_i C_i \\ S_{i,K} & C_{i,K}\cos\alpha_K & C_{i,K}\sin\alpha_K & a_K S_{i,K}+a_{K-1}S_{i,K-1}+\ldots+a_i S_i \\ 0 & \sin\alpha_K & \cos\alpha_K & d_K+d_{K-1}+\ldots+d_i \\ 0 & 0 & 0 & 1 \end{bmatrix}$$

$$(2.2.76)$$

By postmultiplying the above matrix by ^KT_n, we obtain the following recursive relations for the elements of matrices $^{i-1}T_n$, $i=K-1,K-2,\ldots,k$

$$Tn(i-1)1\ell = C_{i,K} TnK1\ell - S_{i,K} TRnK\ell, \qquad \ell=1,2,3 \qquad (2.2.77)$$

$$Tn(i-1)14 = C_{i,K} TPnK - S_{i,K} TRnK4 +$$
$$+ a_{K-1}C_{i,K-1}+a_{K-2}C_{i,K-2}+\ldots+a_{i+1}C_{i,i+1}+a_i C_i \qquad (2.2.78)$$

$$Tn(i-1)2\ell = S_{i,K} TnK1\ell + C_{i,K} TRnK\ell, \qquad \ell=1,2,3 \qquad (2.2.79)$$

$$Tn(i-1)24 = S_{i,K} TPnK + C_{i,K} TRnK4 +$$

$$+ \, a_{K-1}S_{i,K-1} + a_{K-2}S_{i,K-2} + \ldots + a_{i+1}S_{i,i+1} + a_i S_i \qquad (2.2.80)$$

$$Tn(i-1)3\ell = \sin\alpha_K TnK2\ell + \cos\alpha_K TnK3\ell, \qquad \ell=1,2,3 \qquad (2.2.81)$$

$$Tn(i-1)34 = \sin\alpha_K TnK24 + \cos\alpha_K TnK34 +$$

$$+ \, d_K + d_{K-1} + \ldots + d_{i+1} + d_i \qquad (2.2.82)$$

where the variables TRnKℓ, $\ell=1,2,3,4$ and TPnK are given by the same expressions as before

$$TRnK\ell = \cos\alpha_K TnK2\ell - \sin\alpha_K TnK3\ell, \quad \ell=1,2,3,4 \qquad (2.2.83)$$

$$TPnK = TnK14 + a_K \qquad (2.2.84)$$

In the case where K=n, i.e. when at least the two last joints are revolute and parallel, the relations (2.2.77)-(2.2.84) still hold, if the variables Tnnjℓ are taken to correspond to the elements of the (4×4) unit matrix.

It is clear from Equations (2.2.77)-(2.2.82) that the evaluation of the elements of $^{i-1}T_n$ does not require the value of elements of $^{i}T_n$. This implies that the elements of matrix $^{k-1}T_n$ can be obtained according to (2.2.77)-(2.2.82), directly, without evaluating the medial transformations $^{i-1}T_n$, i=K,K-1,...,k+1. This results in a considerably reduced number of floating-point operations. Very often, however, the computation of matrix $^{O}T_n$ is followed by the computation of the Jacobian matrix. In this case it is to be expected that the medial transformation matrices, or at least their fourth columns, are still required. It will be shown in Subsection 2.3.3 that, if the computation of $^{O}T_n$ is followed by the computation of ^{n}J (the Jacobian with respect to the hand coordinate frame), the medial transformations need not in fact be evaluated. Even their fourth columns need not be evaluated, since due to trigonometric compressions it becomes more efficient computationally to evaluate some other variables instead of the sums (2.2.78) and (2.2.80) (see Subsection 2.3.3).

We can summarize thus, that if the computation of $^{O}T_n$ is followed by the computation of ^{n}J, and if the manipulator has revolute joints i=k, k+1,...,K whose axes are parallel the procedure is as follows. First, transformations $^{i-1}T_n$, i=n, n-1,...,K+1 are evaluated, according to the backward recursive symbolic relations for revolute and prismatic joints derived in Subsection 2.2.3. Then, matrix $^{k-1}T_n$ is directly determined, according to (2.2.77)-(2.2.82), as

$$Tn(k-1)1\ell = C_{k,K}TnK1\ell - S_{k,K}TRnK\ell, \qquad \ell=1,2,3$$

$$Tn(k-1)14 = C_{k,K}TPnK-s_{k,K}TRnK4+a_KC_{k,K-1}+...+a_kC_k$$

$$Tn(k-1)2\ell = S_{k,K}TnK1\ell + C_{k,K}TRnK\ell, \qquad \ell=1,2,3 \qquad (2.2.85)$$

$$Tn(k-1)24 = S_{k,K}TPnK+C_{k,K}TRnK4+a_KS_{k,K-1}+...+a_kS_k$$

$$Tn(k-1)3\ell = sin\alpha_K TnK2\ell + cos\alpha_K TnK3\ell, \qquad \ell=1,2,3$$

$$Tn(k-1)34 = sin\alpha_K TnK24+cos\alpha_K TnK34+d_K+d_{K-1}+...+d_k$$

where $TRnK\ell$, $\ell=1,2,3,4$ and $TPnK$ are given by (2.2.83) and (2.2.84). Afterwards, the evaluation of transformations $^{i-1}T_n$, $i=k-1,...,1$ should be continued, according to the backward recursive relations derived in Subsection 2.2.3.

If the Jacobian ^0J with respect to the base reference system is to be computed, one could logically expect that the fourth columns of the medial transformations are necessary, and therefore, should be evaluated from (2.2.78) and (2.2.80). It will be shown, however, in Subsection 2.4.3, that, due to the condensation of trigonometric expressions, it is more appropriate to compute these position vectors in a somewhat different form, which, in fact, corresponds to the terms of the sums for $Tn(k-1)14$ and $Tn(k-1)24$.

Therefore, we will introduce recursive computation of the sums for $Tn(k-1)14$ and $Tn(k-1)24$ in (2.2.85). We will introduce new scalar variables $Tn(k-1)14j$, $j=1,...,K-k$

$$Tn(k-1)141 = C_{k,K}TPnK - S_{k,K}TRnK4$$

$$Tn(k-1)142 = Tn(k-1)141 + a_{K-1}C_{k,K-1}$$
$$\vdots$$
$$Tn(k-1)14j = Tn(k-1)14(j-1)+a_{K-j+1}C_{k,K-j+1}, \qquad 2\leqslant j\leqslant K-k \qquad (2.2.86)$$
$$\vdots$$
$$Tn(k-1)14 = Tn(k-1)14(K-k) + a_kC_k$$

Similarly, we introduce recursive evaluation of the element $Tn(k-1)24$, through new scalar variables $Tn(k-1)24j$, $j=1,\ldots,K-k$

$$Tn(k-1)241 = S_{k,K}TPnK + C_{k,K}TRnK4$$

$$Tn(k-1)242 = Tn(k-1)241 + a_{K-1}S_{k,K-1}$$

$$\vdots \qquad\qquad\qquad\qquad\qquad\qquad\qquad\qquad (2.2.87)$$

$$Tn(k-1)24j = Tn(k-1)24(j-1) + a_{K-j+1}S_{k,K-j+1}, \qquad 2<j<K-k$$

$$\vdots$$

$$Tn(k-1)24 = Tn(k-1)24(K-k) + a_k S_k$$

The variables $Tn(k-1)14j$ and $Tn(k-1)24j$, $j=1,\ldots,K-k$ will have an important role in the evaluation of the Jacobian ^OJ (Subsection 2.4.3). The element $TN(k-1)34$ does not figure in the Jacobian evaluation and can be computed according to (2.2.85).

An illustration of the direct kinematic problem solution for a manipulator with parallel revolute joints, according to the above developed method, will be presented in Appendix I, at the end of this chapter. This example indicates that each pair of parallel joints saves about 10 multiplications and about 5 additions while computing matrix ^OT_n. A rough estimate of the proportionate savings is about $\frac{p}{n}$ 100%, where p denotes the number of pairs of parallel joints, and n is the number of joints. This refers only to the evaluation of matrix ^OT_n by backward recursive relations. However, it will be shown in Subsection 2.3.3 that the total savings are decreased if the computation of ^OT_n is followed by the evaluation of matrix ^nJ. On the other hand, the recursive computation of the sums (2.2.86) and (2.2.87) can be efficiently utilized in evaluating the Jacobian ^OJ with respect to the base reference system, so that the total savings in that case are considerable (see Subsection 2.4.3).

Let us now consider the computation of the homogeneous transformations corresponding to revolute joints with parallel axes, according to the forward recursive relations.

Forward recursive relations

As before, we will denote the serial number of the joint whose axis (axis z_{k-1}) is the first in the series of parallel axes by k, while the last joint with the parallel axis will be designated by K. We will also assume that the axes are adopted to have the same direction, so that $\alpha_i = 0$, $i = k, k+1, \ldots, K-1$ (Fig. 2.6).

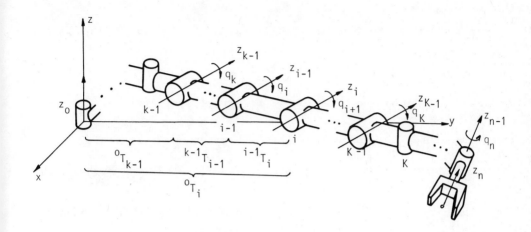

Fig. 2.6. An illustration of the forward recursive relations for joints with parallel axes

By postmultiplying homogeneous transformations, starting from the base out to link k, we can obtain matrices $^0T_1$, $^0T_2, \ldots, ^0T_{k-1}$, ^0T_k. Instead of computing $^0T_{k+1}$ according to $^0T_{k+1} = ^0T_k \, ^kT_{k+1}$, we will first obtain the product of matrices corresponding to parallel joints, and then premultiply the resultant matrix by $^0T_{k-1}$. Accordingly, the recursive relations have the following form

$$^{k-1}T_i = {}^{k-1}T_{i-1} \, {}^{i-1}T_i, \qquad 1 \leqslant k < i \leqslant K \leqslant n \qquad (2.2.88)$$

$$^0T_i = {}^0T_{k-1} \, {}^{k-1}T_i$$

Since the axes z_{k-1}, $z_k, \ldots, z_{i-1}$ are parallel, the transformation matrix $^{k-1}T_i$, $k+1 \leqslant i \leqslant K$ has the following form

$$^{k-1}T_i = \begin{bmatrix} C_{k,i} & -S_{k,i}\cos\alpha_i & S_{k,i}\sin\alpha_i & a_i C_{k,i} + a_{i-1} C_{k,i-1} + \ldots + a_k C_k \\ S_{k,i} & C_{k,i}\cos\alpha_i & C_{k,i}\sin\alpha_i & a_i S_{k,i} + a_{i-1} S_{k,i-1} + \ldots + a_k S_k \\ 0 & \sin\alpha_i & \cos\alpha_i & d_i + d_{i-1} + \ldots d_{k+1} + d_k \\ 0 & 0 & 0 & 1 \end{bmatrix}$$

$$(2.2.89)$$

By premultiplying the above matrix by $^0T_{k-1}$, we obtain the recursive expressions for the elements of matrix ^OT_i, i=k+1, k+2,...,K

$$Tioj1 = C_{k,i}T(k-1)oj1 + S_{k,i}T(k-1)oj2, \qquad j=1,2,3 \qquad (2.2.90)$$

$$Tioj2 = -\cos\alpha_i TSkij + \sin\alpha_i T(k-1)oj3, \qquad j=1,2,3 \qquad (2.2.91)$$

$$Tioj3 = \sin\alpha_i TSkij + \cos\alpha_i T(k-1)oj3, \qquad j=1,2,3 \qquad (2.2.92)$$

$$Tioj4 = a_i Tioj1 + d_i T(k-1)oj3 + T(i-1)oj4, \quad j=1,2,3 \qquad (2.2.93)$$

where, as before, the variables TSkij are given by

$$TSkij = S_{k,i}T(k-1)oj1 - C_{k,i}T(k-1)oj2, \quad j=1,2,3, \; 1 \leqslant k < i \leqslant K \leqslant n \quad (2.2.94)$$

In the case where k=1 the above equations still hold if the elements Toojℓ are taken to correspond to a unit matrix.

Since $\alpha_i = 0$, i=k, k+1,...,K-1 (but not for i=K), the first three columns of the matrices ^OT_i, i=k, k+1,...,K-1 become

$$Tioj1 = C_{k,i}T(k-1)oj1 + S_{k,i}T(k-1)oj2, \qquad j=1,2,3$$

$$Tioj2 = -S_{k,i}T(k-1)oj1 + S_{k,i}T(k-1)oj2, \qquad j=1,2,3 \qquad (2.2.95)$$

$$Tioj3 = T(k-1)oj3, \qquad\qquad j=1,2,3, \quad i=k, \; k+1,...,K-1$$

where for i=k, we assume that $S_{k,k} = S_k$ and $C_{k,k} = C_k$. For the last parallel joint i=K we have

$$TKoj1 = C_{k,K}T(k-1)oj1 + S_{k,K}T(k-1)oj2, \qquad j=1,2,3$$

$$TKoj2 = -\cos\alpha_K TSkKj + \sin\alpha_K T(k-1)oj3, \qquad j=1,2,3 \qquad (2.2.96)$$

$$TKoj3 = \sin\alpha_K TSkKj + \cos\alpha_K T(k-1)oj3, \qquad j=1,2,3$$

where

$$TSkKj = S_{k,K}T(k-1)oj1 - C_{k,K}T(k-1)oj2, \qquad j=1,2,3 \qquad (2.2.97)$$

We will not evaluate the fourth columns of matrices ^OT_i according to (2.2.93), for the reason that they are not directly necessary for the

Jacobian matrix computation. Besides, while the first three columns of ^OT_i do not depend on the elements of $^OT_{i-1}$, this is not the case with the fourth column (see Equations (2.2.90)-(2.2.93)). The position vectors will be evaluated according to the previously derived backward recursive relations (2.2.86), (2.2.87).

It is evident from Equation (2.2.96) that the elements of matrix ^OT_K can be obtained directly, without evaluating the intermediate matrices ^OT_k, $^OT_{k+1}$,..., $^OT_{K-1}$, according to (2.2.95). The Equations (2.2.95) will be utilized only for deriving the expression of the Jacobian ^OJ in Subsection 2.4.3, but will not be evaluated.

An illustration of the application of the forward recursive relations for parallel joints is presented in Appendix I at the end of this chapter. The saving in the number of operations is approximately the same as for the backward recursive evaluation of the transformation matrix ^OT_n.

2.3. The Jacobian Matrix with Respect to the Hand Coordinate Frame

In this section we will consider the problem of automatic, computer--aided generation of the Jacobian matrix, or more precisely, the problem of generating the set of equations which provide for the most computationally efficient evaluation of the Jacobian. In this section we will be concerned with the Jacobian matrix ^nJ which relates between the linear and angular velocities of the manipulator end-effector, expressed with respect to the hand coordinate frame, and joint velocities.

The general expressions for computing this matrix have been derived in Subsection 1.5.2. They can be utilized either for the numeric evaluation of the Jacobian or for writing symbolic expressions for the Jacobian elements by hand given a manipulator. Here, we shall use them in order to develop general symbolic relations which yield the set of equations for optimal evaluation of the Jacobian.

We shall first reconcile the equations derived in Subsection 1.5.2 to the notation used in this chapter, since this is convenient in automatic, computer-aided modelling. We will then demonstrate the manual use of these expressions taking the example of the UMS-3B manipulator. Afterwards, we will derive symbolic expressions for the elements of the

Jacobian, where the compression of trigonometric expressions is performed analitically.

Let us consider the Jacobian matrix ^nJ, i.e.

$$
\begin{bmatrix}
^nv_{nx} \\
^nv_{ny} \\
^nv_{nz} \\
^n\omega_{nx} \\
^n\omega_{ny} \\
^n\omega_{nz}
\end{bmatrix}
=
\begin{bmatrix}
^nJ_c \\
\hline
^nJ_\omega
\end{bmatrix}
\begin{bmatrix}
\dot{q}_1 \\
\vdots \\
\dot{q}_n
\end{bmatrix}
= {}^nJ\dot{q}, \qquad {}^nJ\epsilon R^{6\times n}
\qquad (2.3.1)
$$

wehre $^nv_{nx}$, $^nv_{ny}$, $^nv_{nz}$ are the components of the translational velocity of the manipulator end point in the coordinate system n which is assigned to the last link; $^n\omega_{nx}$, $^n\omega_{ny}$, $^n\omega_{nz}$ – the components of the angular velocity of link n expressed with respect to system n; $^nJ_c\epsilon R^{3\times n}$ and $^nJ_\omega\epsilon R^{3\times n}$ – the matrices given by (1.5.28) and (1.5.45); $\dot{q}\epsilon R^n$ – the vector of joint velocities. The columns of matrices ^nJ_c and $^nJ_\omega$ are given by Equations (1.5.31), (1.5.32) and (1.5.46), in the case of Denavit-Hartenberg kinematic notation. If in these expressions, the matrices are postmultiplied by the vectors, and if the notation (2.2.29) is introduced, column i of matrix ^nJ, $1 < i < n$, which corresponds to a revolute joint, becomes

$$
n_j(i) =
\begin{bmatrix}
-Tn(i-1)11*Tn(i-1)24+Tn(i-1)21*Tn(i-1)14 \\
-Tn(i-1)12*Tn(i-1)24+Tn(i-1)22*Tn(i-1)14 \\
-Tn(i-1)13*Tn(i-1)24+Tn(i-1)23*Tn(i-1)14 \\
Tn(i-1)31 \\
Tn(i-1)32 \\
Tn(i-1)33
\end{bmatrix}
, \quad 1 < i < n \quad (2.3.2)
$$

where * stands for multiplication. If joint i is a sliding one the column of matrix ^nJ is given by

$$
n_j(i) = \begin{bmatrix} Tn(i-1)31 \\ Tn(i-1)32 \\ Tn(i-1)33 \\ 0 \\ 0 \\ 0 \end{bmatrix}, \quad 1 < i < n
$$

(2.3.3)

2.3.1. The Jacobian for the UMS-3B manipulator

In this subsection we will derive analytical expressions for the Jacobian ^nJ of the UMS-3B manipulator (Fig. 2.2) in order to illustrate the complexity of the equations, and to point out the problem of condensing the trigonometric expressions. These analytical forms can also be compared with the computer generated set of equations for efficient evaluation of the Jacobian which will be presented in Subsection 2.3.2.

The homogeneous transformations $^{i-1}T_n$, $i=1,\ldots,n$ for the UMS-3B manipulator have already been derived in Subsection 2.2.2, Equations (2.2.15)--(2.2.21). Since the first joint is a revolute one, the components of the first column of ^nJ are obtained by applying Equation (2.3.2), for $i=1$. Substituting the elements of matrix $^0T_6$ (Eq. (2.2.21)) into (2.3.2) we get

$$
J_{11} = -\{C_1[C_2(C_4C_5C_6+S_4S_6)-S_2S_5C_6]-S_1(S_4C_5C_6-C_4S_6)\}\cdot
$$

$$
\cdot\{S_1\{C_2[C_4C_5(a_6C_6+a_5)+S_4a_6S_6+a_2]-S_2[S_5(a_6C_6+a_5)+q_3]\}+
$$

$$
+ C_1[S_4C_5(a_6C_6+a_5)-C_4a_6S_6]\}+\{S_1[C_2(C_4C_5C_6+S_4S_6)-S_2S_5C_6]+
$$

$$
+ C_1(S_4C_5C_6-C_4S_6)\}\cdot\{C_1\{C_2[C_4C_5(a_6C_6+a_5)+S_4a_6S_6+a_2]-
$$

$$
-S_2[S_5(a_6C_6+a_5)+q_3]\}-S_1[S_4C_5(a_6C_6+a_5)-C_4a_6S_6]\}
$$

(2.3.4)

Upon simplification element J_{11} becomes

$$
J_{11} = -[C_2(C_4C_5C_6+S_4S_6)-S_2S_5C_6][S_4C_5(a_6C_6+a_5)-C_4a_6S_6]+
$$

$$+ \{C_2[C_4C_5(a_6C_6+a_5)+S_4a_6S_6+a_2]-S_2[S_5(a_6C_6+a_5)+q_3]\} \cdot$$

$$\cdot (S_4C_5C_6-C_4S_6) \tag{2.3.5}$$

The other components of the first column of ^nJ may be obtained similarly

$$J_{21} = -\{C_1[C_2(-C_4C_5S_6+S_4C_6)+S_2S_5S_6]-S_1(-S_4C_5S_6-C_4C_6)\}$$

$$\cdot \{S_1\{C_2[C_4C_5(a_6C_6+a_5)+S_4a_6S_6+a_2]-S_2[S_5(a_6C_6+a_5)+q_3]\}+$$

$$+ C_1[S_4C_5(a_6C_6+a_5)-C_4a_6S_6]\}+$$

$$+ \{S_1[C_2(-C_4C_5S_6+S_4C_6)+S_2S_5S_6]+C_1(-S_4C_5S_6-C_4C_6)\} \cdot$$

$$\cdot \{C_1\{C_2[C_4C_5(a_6C_6+a_5)+S_4a_6S_6+a_2]-S_2[S_5(a_6C_6+a_5)+q_3]\}-$$

$$- S_1[S_4C_5(a_6C_6+a_5)-C_4a_6S_6]\} \tag{2.3.6}$$

$$J_{21} = -[C_2(-C_4C_5S_6+S_4C_6)+S_2S_5S_6][S_4C_5(a_6C_6+a_5)-C_4a_6S_6]+$$

$$+ \{C_2[C_4C_5(a_6C_6+a_5)+S_4a_6S_6+a_2]-S_2[S_5(a_6C_6+a_5)+q_3]\} \cdot$$

$$\cdot (-S_4C_5S_6-C_4C_6) \tag{2.3.7}$$

$$J_{31} = -[C_1(C_2C_4S_5+S_2C_5)-S_1S_4S_5] \cdot \{S_1\{C_2[C_4C_5(a_6C_6+a_5)+$$

$$+ S_4a_6S_6+a_2]-S_2[S_5(a_6C_6+a_5)+q_3]\}+C_1[S_4C_5(a_6C_6+a_5)-C_4a_6S_6]\}+$$

$$+ [S_1(C_2C_4S_5+S_2C_5)+C_1S_4S_5] \cdot \{C_1\{C_2[C_4C_5(a_6C_6+a_5)+$$

$$+ S_4a_6S_6+a_2]-S_2[S_5(a_6C_6+a_5)+q_3]\}-S_1[S_4C_5(a_6C_6+a_5)-C_4a_6S_6]\} \tag{2.3.8}$$

$$J_{31} = -(C_2C_4S_5+S_2C_5)[S_4C_5(a_6C_6+a_5)-C_4a_6S_6]+$$

$$+ \{C_2[C_4C_5(a_6C_6+a_5)+S_4a_6S_6+a_2]-S_2[S_5(a_6C_6+a_5)+q_3]\} \cdot S_4S_5$$

$$J_{41} = S_2(C_4C_5C_6 + S_4S_6) + C_2S_5C_6$$

$$\tag{2.3.9}$$

$$J_{51} = S_2(-C_4C_5S_6 + S_4C_6) - C_2S_5S_6$$

$$J_{61} = S_2 C_4 S_5 - C_2 C_5 \qquad (2.3.10)$$

We may observe that none of the components of the first column of $^n J$ depends on joint angle q_1, upon the simplification of the expressions, although the elements of matrix $^0 T_6$ depend on this variable. This is quite logical if one bears in mind the fact that the axis of joint i (z_{i-1} axis) does not move with respect to the coordinate frame n (nor with respect to the base frame) when the joint angle q_i changes its value, and at the same time the position vector from joint i to the manipulator tip, expressed with respect to system n, does not depend on q_i.

The elements of the second column of the Jacobian are again obtained from Equation (2.3.2) and (2.2.21) for i=2. Upon simplification they become

$$J_{12} = -(C_4 C_5 C_6 + S_4 S_6)[S_5(a_6 C_6 + a_5) + q_3] +$$
$$+ [C_4 C_5 (a_6 C_6 + a_5) + S_4 a_6 S_6 + a_2]S_5 C_6$$

$$J_{22} = -(-C_4 C_5 S_6 + S_4 C_6)[S_5(a_6 C_6 + a_5) + q_3] -$$
$$\qquad\qquad (2.3.11)$$
$$- [C_4 C_5 (a_6 C_6 + a_5) + S_4 a_6 S_6 + a_2]S_5 S_6$$

$$J_{32} = -(C_4 S_5)[S_5(a_6 C_6 + a_5) + q_3] -$$
$$- [C_4 C_5 (a_6 C_6 + a_5) + S_4 a_6 S_6 + a_2]C_5$$

$$J_{42} = -S_4 C_5 C_6 + C_4 S_6$$

$$J_{52} = S_4 C_5 S_6 + C_4 C_6$$

$$J_{62} = -S_4 S_5$$

The third joint in the UMS-3B manipulator is sliding, so that the components of the third Jacobian column are obtained from Equation (2.3.3), for i=3, and (2.2.18)

$$J_{31} = S_5 C_6$$
$$J_{32} = -S_5 S_6$$

$$J_{33} = -C_5$$
$$J_{34} = 0$$
$$J_{35} = 0$$
$$J_{36} = 0$$

(2.3.12)

As one works on towards the last degree of freedom, the elements of the Jacobian with respect to the hand coordinate frame become less complex

$$J_{14} = C_5 C_6 a_6 S_6 - C_5 (a_6 C_6 + a_5) S_6 = -C_5 a_5 S_6$$

$$J_{24} = -C_5 S_6 a_6 S_6 - C_5 (a_6 C_6 + a_5) C_6 = -C_5 a_6 - C_5 C_6 a_5$$

$$J_{34} = S_5 a_6 S_6$$

$$J_{44} = S_5 C_6$$

$$J_{54} = -S_5 S_6$$

$$J_{64} = -C_5$$

(2.3.13)

$$J_{15} = 0$$
$$J_{25} = 0$$
$$J_{35} = -a_6 C_6 - a_5$$
$$J_{45} = S_6$$
$$J_{55} = C_6$$
$$J_{65} = 0$$

(2.3.14)

The last column of ^nJ matrix is always a constant column matrix. In the UMS3B manipulator it is given by

$$J_{16} = 0$$
$$J_{26} = 0$$
$$J_{36} = 0$$
$$J_{46} = 0$$
$$J_{56} = 0$$
$$J_{66} = 1.$$

(2.3.15)

2.3.2. Recursive symbolic relations for the Jacobian with respect to the hand coordinate frame

It was noticeable in the previous example that the analitical expressions for the elements of the Jacobian could be compressed and simplified, while this was not the case with the elements of the homogeneous transformation matrices. Since we are interested in the most computationally efficient evaluation of the Jacobian, we must not use Equation (2.3.2) directly, for it does not provide for the minimal number of numerical operations. For example, the evaluation of element J_{15} for the UMS-3B manipulator (Eq. (2.3.14)), according to (2.3.2) and (2.2.16) would yield

$$J_{15} = -T6411*T6424+T6421*T6414 =$$

$$= -(C_5C_6)[S_5(a_6C_6+a_5)]+(S_5C_6)[C_5(a_6C_6+a_5)]$$

Upon simplification if becomes

$$J_{15} = 0$$

We can see that although all the elements of $^4T_6$ needed for the computation of J_{15} are nonzero elements, element J_{15} still equals zero. Thus, the direct application of Equation (2.3.2) would require 2 multiplications and one addition, while the condensed expression requires no numeric operations. We thus conclude that the compression of expressions has generally to be performed analitically in order to obtain the recursive symbolic relations which will yield already condensed expressions. This applies only to the upper three component of the Jacobian columns which correspond to revolute joints, while the other Jacobian elements can not be compressed at all. Namely, these elements coincide with some elements of the homogenous transformations (see Eqs. (2.3.2) and (2.3.3)), whose optimal evaluation has already been considered in Section 2.2.

Let us now derive the condensed symbolic relations for the Jacobian elements. If we assume that joint i is revolute, according to Equation (2.3.2) the upper three elements of column i, i.e. J_{1i}, J_{2i} and J_{3i} can be represented in the form of a single recursive relation

$$J_{\ell i} = -Tn(i-1)1\ell*Tn(i-1)24+Tn(i-1)2\ell*Tn(i-1)14, \quad \ell=1,2,3 \quad (2.3.16)$$

By substituting the recursive expression (2.2.39)-(2.2.42) for the elements of the homogeneous transformations, we obtain

$$J_{\ell i} = -[cosq_i Tni1\ell - sinq_i TRni\ell][sinq_i TPni + cosq_i TRni4] +$$

$$+ [sinq_i Tni1\ell + cosq_i TRni\ell][cosq_i TPni - sinq_i TRni4]$$

Upon simplification this becomes

$$J_{\ell i} = -Tni1\ell * TRni4 + TPni * TRni\ell, \qquad \ell = 1,2,3, \qquad 1 \leqslant i \leqslant n \qquad (2.3.17)$$

All the variables appearing in the above expression have already been evaluated during the computation of matrix $^{O}T_n$ according to the backward recursive algorithm presented in Subsection 2.2.3. We can see, that generally, once the direct problem has been solved, the number of multiplications and additions in Equation (2.3.17) is the same as in (2.3.16). However, since the variables in Equation (2.3.17) are functions of one joint coordinate less than the variables in (2.3.16), they are automatically simpler. In case where some variables from Equation (2.3.17) happen to be 0 or ±1, the number of multiplications and additions is reduced from 2 and 1 up to 0. On the other hand, if the element was evaluated according to (2.3.16) it probably would not happen (as in the example of element J_{15}).

Besides the above described compression, which occurs regularly with every Jacobian column which corresponds to a revolute joint, it may happen that the trigonometric expressions can be further condensed. Let us, for example, consider the evaluation of element J_{14} given by Eq. (2.3.13). Applying relation (2.3.16) to the homogeneous transformation $^{3}T_6$ given by (2.2.17), we obtain

$$J_{14} = -(C_4 C_5 C_6 + S_4 S_6)[S_4 C_5(a_6 C_6 + a_5) - C_4 a_6 S_6] +$$

$$+ (S_4 C_5 C_6 - C_4 S_6)[C_4 C_5(a_6 C_6 + a_5) + S_4 a_6 S_6]$$

Upon eliminating angle q_4 we obtain

$$J_{14} = (C_5 C_6)(a_6 S_6) - [C_5(a_6 C_6 + a_5)]S_6$$

this expression would also be obtained from (2.3.17), and would then require two multiplications and one subtraction. However, the expression can be further compressed into

$$J_{14} = -(C_5 S_6) a_5$$

which requires only one multiplication and no additions, since the product $C_5 S_6$ has already been evaluated while computing $^0 T_n$.

However, an example of the opposite is element J_{24} (Eq. (2.3.13)), where the number of operations does not decrease although further compressing is possible

$$J_{24} = -C_5 S_6 a_6 S_6 - C_5 (a_6 C_6 + a_5) C_6 = -C_5 a_6 - (C_5 C_6) a_6$$

With other elements of the Jacobian for the UMS-3B manipulator, such as J_{11}, J_{21}, J_{31}, J_{12}, J_{22}, J_{32}, J_{13}, J_{23} and J_{33}, upon multiplying the expressions in parantheses and canceling some terms, the number of operations would become considerably larger. This is due to the fact that in this case the variables would be, grouped in a different manner, different from the previous one. For example, the element J_{12}, would require two multiplications and one addition if evaluated according to (2.3.17). However, upon multiplication and compression it becomes

$$J_{12} = -(S_4 S_6) S_5 - (C_4 C_5 C_6 + S_4 S_6) q_3 + a_2 (S_5 C_6)$$

This expression now requires 3 multiplications and 2 additions, since the terms in parantheses have been determined while computing $^0 T_n$. The more complex the terms in (2.3.17), the larger the number of operations would be upon their disassembling.

As we have seen from the above examples, the variables in (2.3.17) could be expressed in terms of the elements of matrix $^{i+1} T_n$. However, this would result in very modest computational savings in the evaluation of a very small number of Jacobian elements (with the UMS-3B manipulator it is only the element J_{14}). On the other hand, this would be a considerable burden on the off-line algorithm for automatic computer-aided evaluation of the Jacobian so we will no longer take it into consideration.

As has been already mentioned, the lower three elements of the Jacobian columns which correspond to revolute joints, require no additional floating-point operations, since they coincide with the third rows of homogeneous transformations $^{i-1} T_n$. According to Equation (2.3.2) we can write the recursive relation directly

$$J_{\ell i} = Tn(i-1)3(\ell-3), \qquad \ell=4,5,6, \quad 1 \leqslant i \leqslant n \qquad (2.3.18)$$

The elements of the Jacobian columns which correspond to sliding joints are given by Equation (2.3.3), which can be presented in the equivalent form

$$J_{\ell i} = Tn(i-1)3\ell, \qquad \ell=1,2,3$$
$$1 \leqslant i \leqslant n \qquad (2.3.19)$$
$$J_{\ell i} = 0, \qquad \ell=1,2,3$$

Thus, all the recursive expressions required for automatic generation of the set of equations for optimal evaluation of the Jacobian matrix with respect to the hand coordinate frame, have been derived.

Let us illustrate the evaluation of the Jacobian matrix ^nJ for the UMS-3B manipulator according to recursive expressions (2.3.17)-(2.3.19). While writing these equations we will, as before, avoid adding zero variables, multiplying by ±1, and in general, replace all the simple variables by their real contents (see the definition of simple and complex variables in Subsection 2.2.3).

Since the first two degrees of freedom of the UMS-3B manipulator (Fig. 2.2) are revolute, the elements of the first and second column are obtained from (2.3.17), (2.3.18). We will also make use of the set of Equations (2.2.49)-(2.2.55), which evaluate the matrices $^{i-1}T_6$, $i=6,...$ ...,1 for this manipulator (derived in Subsection 2.2.3). So, if we designate by $J\ell i$ the elements $J_{\ell i}$, $\ell=1,...,6$, $i=1,...,6$, we obtain

$$J11 = -T6111*T6324+T6114*T6321$$
$$J21 = -T6112*T6324+T6114*T6322$$
$$J31 = -T6113*T6324+T6114*T6323$$
$$J41 = T6121$$
$$J51 = T6122$$
$$J61 = T6123 \qquad (2.3.20)$$
$$J12 = -T6311*T6234+TP62*T6421$$
$$J22 = -T6312*T6234+TP62*T6422$$
$$J32 = -T6313*T6234-TP62*C_5$$
$$J42 = -T6321$$

$$J52 = -T6322$$

$$J62 = -T6323$$

The third column is obtained from (2.3.19) as

$$J13 = T6421$$

$$J23 = T6422$$

$$J33 = -C_5$$

$$J43 = 0.$$

$$J53 = 0.$$

$$J63 = 0.$$

(2.3.21)

The right three columns of the Jacobian with respect to the hand co-ordinate system, are otpimally evaluated from the following set of equations

$$J14 = T6411*T6524-T6414*S_6$$

$$J24 = T6412*T6524-T6414*C_6$$

$$J34 = S_5*T6524$$

$$J44 = T6421$$

$$J54 = T6422$$

$$J64 = -C_5$$

$$J15 = 0.$$

$$J25 = 0.$$

$$J35 = -TP65$$

$$J45 = S_6$$

$$J55 = C_6$$

$$J65 = 0.$$

$$J16 = 0.$$

$$J26 = 0.$$

$$J36 = 0.$$

$$J46 = 0.$$

$$J56 = 0.$$

$$J66 = 1.$$

(2.3.22)

It follows from the above equations, that the computation of the Jacobian ^nJ for the UMS-3B manipulator requires an additional 17 multiplications and 8 additions, once the elements of ^oT_n have been determined. Since the direct kinematic problem required 54 multiplications, 25 additions and 10 transcedental function calls (see Subsection 2.2.3), the total number of operations required for computing ^oT_n an ^nJ is 71 multiplications, 33 additions and 10 transedental function calls, in the UMS-3B manipulator. Let us remark here that this manipulator belongs to a class of industrial robots with average kinematic complexity. A more detailed discussion on numerical complexity of the computer generated symbolic kinematic model in various manipulator configurations, will be given in Section 2.5.

The above example showed that the evaluation of matrix ^nJ required the evaluation of all the elements of matrices $^5T_6$, $^4T_6$, $^3T_6$, $^2T_6$ and $^1T_6$, except for the element T6124. The matrix $^oT_6$ is not necessary for the evaluation of the Jacobian with respect to the hand coordinate frame. Therefore, if only the evaluation of ^nJ is desired, the set of equations is obtained by simply rejecting those equations from the derived set, evaluating the unnecessary variables.

2.3.3. The Jacobian columns corresponding to parallel joints

In the case where the manipulator has at least two revolute joints whose axes are always parallel, the evaluation of the Jacobian matrix should be further optimized. In Subsection 2.2.5 we showed that the evaluation of matrix ^oT_n did not require the evaluation of the intermediate matrices corresponding to parallel joints. In this subsection we will try to derive the recursive expressions for computing the Jacobian columns which require neither the evaluation of the intermediate matrices, nor their fourth columns. Instead, some other sums representing the condensed expressions for the Jacobian elements will be evaluated recursively, providing for the reduced computational complexity.

We will assume, as before, that the axes z_{i-1} of the revolute joints i, i=k, k+1,...,K, are parallel, $1 \leqslant k < K \leqslant n$. First, we shall consider the evaluation of the upper three elements of columns $^nj^{(i)}$ of matrix ^nJ, i=k, k+1,...,K-1. To obtain the recursive expressions we will make use of the expressions (2.2.70)-(2.2.80) specially derived for the elements of $^{i-1}T_n$ for parallel joints. Substituting these equations in the general expression (2.3.2), the upper three elements of column $n_j(i)$ are thus obtained

$$J_{\ell i} = -(C_{i,K}TnK1\ell - S_{i,K}TRnK\ell)(S_{i,K}TPnK + C_{i,K}TRnK4 +$$

$$+ a_{K-1}S_{i,K-1} + a_{K-2}S_{i,K-2} + \cdots + a_{i+1}S_{i,i+1} + a_iS_i) +$$

$$+ (S_{i,K}TnK1\ell + C_{i,K}TRnK\ell)(C_{i,K}TPnK - S_{i,K}TRnK4 +$$

$$+ a_{K-1}C_{i,K-1} + a_{K-2}C_{i,K-2} + \cdots + a_{i+1}C_{i,i+1} + a_iC_i), \quad \ell=1,2,3,$$

$$i=k, \ k+1,\ldots,K-1 \quad (2.3.23)$$

Upon simplification, using the identities

$$S_{i,K}C_{i,j} - C_{i,K}S_{i,j} = S_{j+1,K}$$
$$i < j < K-1$$
$$S_{i,K}S_{i,j} + C_{i,K}C_{i,j} = C_{j+1,K}$$

we obtain

$$J_{\ell i} = TRnK\ell(TPnK + a_{K-1}C_K + a_{K-2}C_{K-1,K} + \cdots + a_{i+1}C_{i+2,K} + a_iC_{i+1,K}) +$$

$$+ TnK1\ell(-TRnK4 + a_{K-1}S_K + a_{K-2}S_{K-1,K} + \cdots + a_{i+1}S_{i+2,K} + a_iS_{i+1,i}),$$

$$\ell=1,2,3, \quad i=k, \ k+1,\ldots,K-1 \quad (2.3.24)$$

It is clear that the above sums can be computed recursively. We will now introduce the new scalar variables JPj, $j=1,\ldots,K-i$ in order to evaluate the first sum

$$JP1 = TPnK + a_{K-1}C_K$$

$$JP2 = JP1 + a_{K-2}C_{K-1,K}$$
$$\vdots$$
$$JPj = JP(j-1) + a_{K-j}C_{K-j+1,K} \ , \quad 1 < j < K-i, \quad k < i < K-1$$
$$\vdots$$
$$JP(K-i) = JP(K-i-1) + a_iC_{i+1,K}$$

$$(2.3.25)$$

Similarly, the second sum is evaluated as

$$JR1 = -TRnK4 + a_{K-1}S_K$$

$$JR2 = JR1 + a_{K-2}S_{K-1,K}$$

$$\vdots \qquad\qquad\qquad\qquad\qquad\qquad\qquad (2.3.26)$$

$$JRj = JR(j-1) + a_{K-2}S_{K-j+1,K} \quad , \quad 1 \leqslant j \leqslant K-i, \quad k \leqslant i \leqslant K-1$$

$$\vdots$$

$$JR(K-i) = JR(K-i+1) + a_i S_{i+1,K}$$

Once all the variables JPj and JRj, j=1,...,K-k are evaluated, the upper three elements of the Jacobian columns $n_j(i)$, i=K-1, K-2,...,k, are obtained from

$$J_{\ell i} = TRnK\ell*JP(K-i) + TnK1\ell*JR(K-i), \quad \ell=1,2,3, \quad i=K-1, K-2,...,k$$

$$(2.3.27)$$

This expression does not really require the evaluation of intermediate transformations $^{i-1}T_n$, i=k+1,...,K. The same is true for the lower three elements of Jacobian column $n_j(i)$, i=k, k+1,...,K-1. Namely, according to (2.3.18)

$$J_{\ell i} = Tn(i-1)3(\ell-3), \quad \ell=4,5,6$$

On the other hand, the third rows of the intermediate transformations are constant, and according to (2.2.81) and (2.2.85) are given by

$$Tn(i-1)3\ell = Tn(k-1)3\ell = \sin\alpha_K TnK2\ell + \cos\alpha_K TnK3\ell, \quad \ell=1,2,3$$

Thus, the evaluation of the lower three elements of Jacobian columns $n_j(i)$, i=k, k+1,...,K-1 is carried out according to

$$J_{\ell i} = Tn(k-1)3(\ell-3), \quad \ell=4,5,6, \quad i=k, k+1,...,K-1 \qquad (2.3.28)$$

requiring no additional numerical operations.

An illustration of the above method of computing the Jacobian matrix with respect to the hand coordinate system, will be given in Appendix II, at the end of this chapter. Clearly, the evaluation of the sums (2.3.25) and (2.3.26) increases the number of operations needed for the Jacobian matrix computation, with respect to the case in which the Jacobian columns were calculated as for ordinary revolute joints. However, this enabled us to omit the evaluation of the intermediate transformations thus gaining greater computational savings. In the example shown in Appendix II the total savings, taking into account the evaluation of both the matrices ${}^{o}T_n$ and ${}^{n}J$, are about 20% of the total number of operations required if the joints were treated as ordinary revolute joints.

2.4. The Jacobian Matrix with Respect to the Base Coordinate Frame

In this section we will deal with the evaluation of the Jacobian matrix relating between linear and angular velocity of the manipulator end-effector, expressed in the base reference coordinate system, and the joint rates. We will first transfer the expressions derived in Subsection 1.5.2 to the notation adopted in this chapter, which is convenient for automatic symbolic modelling. We will illustrate the manual use of these expressions by deriving the analytical expression for the Jacobian ${}^{o}J$ for the UMS-3B manipulator. Then, the recursive relations which are appropriate for computer-aided, computationally optimal evaluation of this matrix, will be derived. The computation of the Jacobian columns corresponding to revolute joints with parallel joint axes, will be considered separately.

The Jacobian matrix ${}^{o}J$ is defined by the following equation

$$
\begin{bmatrix}
{}^{o}v_{nx} \\
{}^{o}v_{ny} \\
{}^{o}v_{nz} \\
{}^{o}\omega_{nx} \\
{}^{o}\omega_{ny} \\
{}^{o}\omega_{nz}
\end{bmatrix}
=
\begin{bmatrix}
{}^{o}J_c \\
\hdashline
{}^{o}J_\omega
\end{bmatrix}
\begin{bmatrix}
\dot{q}_1 \\
\vdots \\
\dot{q}_n
\end{bmatrix}
= {}^{o}J\dot{q}, \qquad {}^{o}J \in R^{6 \times n} \qquad (2.4.1)
$$

where ${}^{o}v_{nx} = \dot{x}$, ${}^{o}v_{ny} = \dot{y}$, ${}^{o}v_{nz} = \dot{z}$ are the components of end-effector translational velocity expressed in the reference coordinate system attached to the manipulator base; ${}^{o}\omega_{nx}$, ${}^{o}\omega_{ny}$, ${}^{o}\omega_{nz}$ - the projections of the angular velocity of link n onto the axes of the base frame; ${}^{o}J_c \in R^{3 \times n}$, ${}^{o}J_\omega \in R^{3 \times n}$ are Jacobian submatrices given by (1.5.22) and (1.5.41); $\dot{q} \in R^n$ is the vector of joint velocities. The columns of matrices ${}^{o}J_c$ and ${}^{o}J_\omega$ are given by Equations (1.5.26) and (1.5.42), in cases where Denavit-Hartenberg kinematic notation is used. If in these equations, the matrices are postmultiplied by the vectors, and if the notation (2.2.29) is introduced, column i of matrix $^O J$, $1 \leqslant i \leqslant n$, which corresponds to a revolute joint, becomes

$$
o_j(i) = \begin{bmatrix}
-T(i-1)011*Tn(i-1)24+T(i-1)012*Tn(i-1)14 \\
-T(i-1)021*Tn(i-1)24+T(i-1)022*Tn(i-1)14 \\
-T(i-1)031*Tn(i-1)24+T(i-1)032*Tn(i-1)14 \\
T(i-1)013 \\
T(i-1)023 \\
T(i-1)033
\end{bmatrix}, \quad 1 \leqslant i \leqslant n \quad (2.4.2)
$$

The columns of matrix $^O J$, which correspond to prismatic joints, are given by

$$
o_j(i) = \begin{bmatrix}
T(i-1)013 \\
T(i-1)023 \\
T(i-1)033 \\
0 \\
0 \\
0
\end{bmatrix}, \quad 1 \leqslant i \leqslant n \quad (2.4.3)
$$

For i=1, one should take that the elements T00jℓ correspond to the (4×4) unit matrix.

2.4.1. The Jacobian for the UMS-3B manipulator

To illustrate the application of the above equations, we will derive analytical expressions for the elements of Jacobian matrix $^O J$ for the UMS-3B manipulator shown in Fig. 2.2. We will also make use of this example to point out the problem of compressing the trigonometric expressions.

The homogeneous transformation matrices $^O T_i$, i=1,...,6 for this manipulator have been determined in Subsection 2.2.2, Equations (2.2.22)-(2.2.28). The fourth columns of matrices $^{i-1} T_n$, i=1,...,6 necessary for the generation of matrix ^{O}J according to (2.4.2), (2.4.3), have also been derived in this subsection.

We apply Equation (2.4.2) in order to obtain expressions for the elements of the first two columns

$$J_{11} = -S_1\{C_2[C_4C_5(a_6C_6+a_5)+S_4a_6S_6+a_2]-S_2[S_5(a_6C_6+a_5)+q_3]\}-$$
$$-C_1[S_4C_5(a_6C_6+a_5)-C_4a_6S_6]$$

$$J_{21} = C_1\{C_2[C_4C_5(a_6C_6+a_5)+S_4a_6S_6+a_2]-S_2[S_5(a_6C_6+a_5)+q_3]\}-$$
$$- S_1[S_4C_5(a_6C_6+a_5)-C_4a_6S_6]$$

$$(2.4.4)$$

$$J_{31} = 0.$$

$$J_{41} = 0.$$

$$J_{51} = 0.$$

$$J_{61} = 1.$$

$$J_{12} = -C_1\{S_2[C_4C_5(a_6C_6+a_5)+S_4a_6S_6+a_2]+C_2[S_5(a_6C_6+a_5)+q_3]\}$$

$$J_{22} = -S_1\{S_2[C_4C_5(a_6C_6+a_5)+S_4a_6S_6+a_2]+C_2[S_5(a_6C_6+a_5)+q_3]\}$$

$$J_{32} = C_2[C_4C_5(a_6S_6+a_5)+S_4a_6S_6+a_2]-S_2[S_5(a_6C_6+a_5)+q_3]$$

$$(2.4.5)$$

$$J_{42} = S_1$$

$$J_{52} = -C_1$$

$$J_{62} = 0.$$

The third degree of freedom is sliding. Therefore the third column is obtained from (2.4.3) and (2.2.23)

$$J_{13} = -S_2C_1$$

$$J_{23} = -S_2S_1$$

$$J_{33} = C_2$$

$$(2.4.6)$$

$$J_{43} = 0$$

$$J_{53} = 0.$$

$$J_{63} = 0.$$

The last three columns correspond again to revolute joints

$$J_{14} = -C_2 C_1 [S_4 C_5 (a_6 C_6 + a_5) - C_4 a_6 S_6] - S_1 [C_4 C_5 (a_6 C_6 + a_5) + S_4 a_6 S_6]$$

$$J_{24} = -C_2 S_1 [S_4 C_5 (a_6 C_6 + a_5) - C_4 a_6 S_6] + C_1 [C_4 C_5 (a_6 C_6 + a_5) + S_4 a_6 S_6]$$

$$J_{34} = -S_2 [S_4 C_5 (a_6 C_6 + a_5) - C_4 a_6 S_6]$$

$$\text{(2.4.7)}$$

$$J_{44} = -S_2 C_1$$

$$J_{54} = -S_2 S_1$$

$$J_{64} = C_2$$

$$J_{15} = -(C_4 C_2 C_1 - S_4 S_1) S_5 (a_6 C_6 + a_5) - S_2 C_1 C_5 (a_6 C_6 + a_5)$$

$$J_{25} = -(C_4 C_2 S_1 + S_4 C_1) S_5 (a_6 C_6 + a_5) - S_2 S_1 C_5 (a_6 C_6 + a_5)$$

$$J_{35} = -C_4 S_2 S_5 (a_6 C_6 + a_5) + C_2 C_5 (a_6 C_6 + a_5)$$

$$\text{(2.4.8)}$$

$$J_{45} = S_4 C_2 C_1 + C_4 S_1$$

$$J_{55} = S_4 C_2 S_1 - C_4 C_1$$

$$J_{65} = S_4 S_2$$

$$J_{16} = -[C_5 (C_4 C_2 C_1 - S_4 S_1) - S_5 S_2 C_1] a_6 S_6 + (S_4 C_2 C_1 + C_4 S_1) a_6 C_6$$

$$J_{26} = -[C_5 (C_4 C_2 S_1 + S_4 C_1) - S_5 S_2 S_1] a_6 S_6 + (S_4 C_2 S_1 - C_4 C_1) a_6 C_6$$

$$J_{36} = -(C_5 C_4 S_2 + S_5 C_2) a_6 S_6 + S_4 S_2 a_6 C_6$$

$$\text{(2.4.9)}$$

$$J_{46} = S_5 (C_4 C_2 C_1 - S_4 S_1) + C_5 S_2 C_1$$

$$J_{56} = S_5 (C_4 C_2 S_1 + S_4 C_1) + C_5 S_2 C_1$$

$$J_{66} = S_5 C_4 S_2 - C_5 C_2$$

2.4.2. Recursive symbolic relations for the Jacobian
with respect to the base coordinate frame

The above example shows that all the columns in matrix ^OJ depend on all the joint coordinates $q_1,\ldots,q_n$, as opposed to the columns of matrix ^nJ, where column i depends solely on coordinates $q_{i+1},\ldots,q_n$. We may observe that the problem of condensing the analytical expressions did not appear in the above example. This is normal, since the elements of the rotation matrix $^OA_{i-1}$ depend only on coordinates $q_1, q_2,\ldots,q_{i-1}$, while the fourth columns of matrix $^{i-1}T_n$ is the function of the remaining coordinates $q_i, q_{i+1},\ldots,q_n$. They can not be compressed at all after multiplication. For this reason this method of evaluating the Jacobian matrix, by combining the use of matrices ^OT_i and $^{i-1}T_n$ according to (2.4.2) and (2.4.3), is very convenient for automatic, computer--aided evaluation of the symbolic Jacobian matrix. We will represent them in a form analogous to the previously developed recursive relations. Thus the elements of the Jacobian columns $^Oj^{(i)}$, $1 \leqslant i \leqslant n$, which correspond to rotational joints (Eq. 2.4.2) are optimally evaluated from

$$J_{\ell i} = -T(i-1)0\ell1*Tn(i-1)24+T(i-1)0\ell2*Tn(i-1)14,$$

$$\ell=1,2,3, \qquad 1\leqslant i \leqslant n \qquad (2.4.10)$$

$$J_{\ell i} = T(i-1)0(\ell-3)3, \qquad \ell=4,5,6, \qquad 1\leqslant i \leqslant n \qquad (2.4.11)$$

For sliding degrees of freedom, the Jacobian columns are

$$J_{\ell i} = T(i-1)0\ell3, \qquad \ell=1,2,3, \qquad 1\leqslant i \leqslant n \qquad (2.4.12)$$

$$J_{\ell i} = 0, \qquad \ell=4,5,6, \qquad 1\leqslant i \leqslant n \qquad (2.4.13)$$

The evaluation of matrix ^OJ according to the above equations, clearly requires the computation of the left three columns of matrices $^OT_{i-1}$, $i=2,\ldots,n$ by forward recursive relations, and the evaluation of the fourth columns of $^{i-1}T_n$, $i=1,\ldots,n$ by backward recursive relations.

On the other hand, if we prefer to employ forward recursive relations in evaluating the Jacobian with respect to the base coordinate system only, the position vectors $^O\vec{p}_i$ between joint i and manipulator tip should be expressed as the difference between the fourth columns of matrices ^OT_n and $^OT_{i-1}$. These vectors, expressed in the base system, are functions of all the joint coordinates, while the vector of joint axis z_{i-1} expressed with respect to the base system depends on $q_1,\ldots$ $\ldots,q_{i-1}$. The vector cross product of these variables can undoubtedly be compressed. In a general case, it is hard to take this compression into account analytically, in advance in a general case. On the other hand, the result of the compression must be the same as the result obtained from (2.4.10). We conclude that the optimal evaluation of the Jacobian with respect to the base reference system, requires the combination of backward and forward recursive relations.

An example of the application of the derived recursive relations (2.4.10)- -(2.4.13) will be given in Section 2.5 in the form of a computer program for evaluating the $^OT_6$ matrix and the Jacobian ^OJ for the UMS-3B manipulator.

2.4.3. The Jacobian columns corresponding to parallel joints

In the case when a manipulator has revolute joints with parallel joint axes it is still possible to optimize the evaluation of the Jacobian with respect to the base coordinate system. This optimization should ensure that the itermediate transformation matrices which correspond to the parallel joints need not be evaluated, as it was the case with the evaluation of the ^OT_n matrix, thus increasing the speed of computation. This will be achieved by a new rearrangment of the variables which correspond to parallel joints. Naturally, this compression does not eliminate any joint coordinates from the expressions of the Jacobian elements, but reduces computational complexity.

We will assume, as before, that the joint axes z_{i-1}, $i=k,\ k+1,\ldots,K$ are parallel, $1<k<K<n$. The upper three elements of column $^Oj^{(i)}$ are given by (2.4.10) if the joint is a revolute one. On the other hand, the elements of $^OT_{i-1}$, $i=k+1,\ldots,K$ were derived in Subsection 2.2.5 (Equations (2.2.95)), while the fourth columns of $^{i-1}T_n$ matrices, $i=k$, $k+1,\ldots,K-1$, are given by (2.2.78) and (2.2.80). By substituting these expressions into (2.4.10) we obtain

$$J_{\ell i} = -(C_{k,i-1}T(k-1)0\ell1 + S_{k,i-1}T(k-1)0\ell2)(S_{i,K}TPnK + C_{i,K}TRnK4 +$$

$$+a_{K-1}S_{i,K-1} + a_{K-2}S_{i,K-2} + \ldots + a_{i+1}S_{i,i+1} + a_iS_i) +$$

$$+(-S_{k,i-1}T(k-1)0\ell1+C_{k,i-1}T(k-1)0\ell2)(C_{i,K}TPnK-S_{i,K}TRnK4 +$$

$$+a_{K-1}C_{i,K-1}+a_{K-2}C_{i,K-2}+\cdots+a_{i+1}C_{i,i+1}+a_iC_i),$$

$$\ell=1,2,3, \qquad i=k+1,\ldots,K-1 \quad (2.4.14)$$

Upon multiplication, by using the following identities

$$C_{k,i-1}S_{i,K-j} + S_{k,i-1}C_{i,K-j} = S_{k,K-j}$$

$$0<j<K-i$$

$$C_{k,i-1}C_{i,K-j} - S_{k,i-1}S_{i,K-j} = C_{k,K-j}$$

we obtain

$$J_{\ell i} = -T(k-1)0\ell1(S_{k,K}TPnK+C_{k,K}TRnK4+a_{K-1}S_{k,K-1} +$$

$$+a_{K-2}S_{k,K-2}+\cdots+a_{i+1}S_{k,i+1}+a_iS_{k,i}) +$$

$$+T(k-1)0\ell2(C_{k,K}TPnK-S_{k,K}TRnK4+a_{K-1}C_{k,K-1} +$$

$$+a_{K-2}C_{k,K-2}+\cdots+a_{i+1}C_{k,i+1}+a_iC_{k,i}),$$

$$\ell=1,2,3 \qquad i=k+1,\ldots,K-1 \quad (2.4.15)$$

If i=K, the Equation (2.4.14) becomes

$$J_{\ell i} = -(C_{k,K-1}T(k-1)0\ell1+S_{k,K-1}T(k-1)0\ell2)(S_KTPnK+C_KTRnK4) +$$

$$+(-S_{k,K-1}T(k-1)0\ell1+C_{k,K-1}T(k-1)0\ell2)(C_KTPnK-S_KTRnK4),$$

$$\ell=1,2,3 \qquad (2.4.16)$$

Upon simplification it becomes

$$J_{\ell i} = -T(k-1)0\ell1(S_{k,K}TPnK+C_{k,K}TRnK4) +$$

$$+ T(k-1)0\ell2(C_{k,K}TPnK-S_{k,K}TRnK4), \qquad \ell=1,2,3 \quad (2.4.17)$$

We can see from Equations (2.4.15) and (2.4.17) that the sums in the parenthesis were already recursively evaluated during the computation of the position vector of matrix $^{k-1}T_n$, according to (2.2.86) and (2.2.87). Therefore, we can now represent the Equations (2.4.15) and (2.4.17) in the following recursive form

$$J_{\ell i} = -T(k-1)0\ell 1*Tn(k-1)24(K-i+1)+T(k-1)0\ell 2*Tn(k-1)14(K-i+1),$$

$$\ell=1,2,3, \quad i=k+1,\ldots,K-1,K \qquad (2.4.18)$$

For i=k, we have

$$J_{\ell i} = -T(k-1)0\ell 1*Tn(k-1)24+T(k-1)0\ell 2*Tn(k-1)14$$

which is the same result as would be obtained by applying the Equation (2.4.10) for i=k.

The recursive relation (2.4.18) gives the upper three elements of the Jacobian matrix columns which correspond to parallel joints. We can see that this expresson also does not make use of the elements of the intermediate matrices corresponding to parallel joints. Thus, computational complexity is considerably reduced.

Neither does the evaluation of the lower three elements of columns $o_j(i)$, i=k+1,...,K require the evaluation of the intermediate matrices. Namely, according to (2.4.11), these elements are given by

$$J_{\ell i} = T(i-1)0(\ell-3)3, \qquad \ell=4,5,6$$

On the other hand, the third columns of the intermediate transformations are constant, and according to (2.2.95) are given by

$$T(i-1)0\ell 3 = T(k-1)0\ell 3, \qquad \ell=1,2,3, \quad i=k+1,\ldots,K$$

Thus, the evaluation of the lower three elements of the Jacobian columns $o_j(i)$, i=k+1,...,K requires no additional floating-point operations

$$J_{\ell i} = T(k-1)0(\ell-3)3, \qquad \ell=4,5,6, \quad i=k+1,\ldots,K \qquad (2.4.19)$$

An illustration of the above method of computing the Jacobian matrix with respect to the base reference system will be given in Appendix III, at the end of this chapter. This example shows that the computational complexity is considerably reduced if special treatment of parallel joints is introduced into the evaluation technique. For the 6 degree--of-freedom arthropoid manipulator with three parallel revolute joints, computational complexity is reduced by about 50%, when the evaluation of both the Jacobian with respect to the base system and the direct kinematic problem are considered. This is mainly due to the facts that the intermediate transformations were not evaluated, and that the evaluation of the Jacobian according to (2.4.18) - (2.4.19) required exclusively the terms which were determined during the solution of the direct kinematic problem.

2.5. Programm Implementation, Numerical Aspects and Examples

In the preceding sections the mathematical background was presented to ensure the computationally optimal evaluation of the symbolic kinematic model of an arbitrary serial link manipulator. In this section we will describe the program realization of the algorithm by which the generation of the equations is completely automatized. The input of the program are Denavit-Hartenberg kinematic parameters of the mechanism, while the output is a computer program which evaluates the kinematic model of the manipulator. Several examples of computer-generated programs will be given. A comparative analysis of the computational complexity of the kinematic models of various robots will be also discussed. The execution times measured on minicomputer PDP 11/70 and a microcomputer based on microprocessor INTEL 8086 with numerical coprocessor 8087, will be given.

2.5.1. Block-diagram of the program for the symbolic model generation

As we have seen in the previous sections, the optimal evaluation of the kinematic variables is carried out according to the derived recursive symbolic relations. The step from these formulae to the computer program evaluating matrix $^{O}T_n$ and the Jacobian is a simple one. One has only to automatize the method of applying these relations which has been illustrated in several examples in the preceding sections. A global block-diagram of the program package which represents the program implementation of the algorithm is shown in Fig. 2.7.

The input of this program package represents the Denavit-Hartenberg kinematic parameters of the manipulator whose model is to be obtained, these being: α_i, a_i, d_i (for revolute joints), θ_i (for sliding joints), ξ_i (indicator as to whether joint i is revolute $\xi_i=0$, or sliding $\xi_i=1$). The output is the "expert" program written in a desired programming language. Given the numeric values of parameters and joint coordinates,

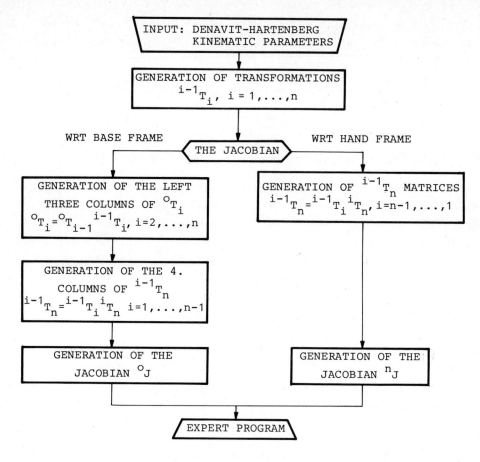

Fig. 2.7. Block-diagram of the program for generating the expert
program for evaluating the kinematic model

the expert program, upon compilation, evaluates the numeric values of
the manipulator hand position, the orientation matrix and the Jacobian
matrix.

The program package for the kinematic model generation starts with the
generation of transformation matrices $^{i-1}T_i$, i=1,...,n between adjacent
link coordinate frames. If the evaluation of a variable requires at
least one additional floating-point operation, the variable is considered as complex and a program instruction to evaluate the variable is
written into the output file. If a variable is a simple one (requires
no additional floating-point operations), nothing is written into the
output file, but the content of the variables is stored in the computer
memory. This procedure eliminates unnecessary shifting of the variables
from one into another.

Upon forming the adjacent transformations $^{i-1}T_i$, the operator running the program has to choose which Jacobian matrix needs to be generated. If the Jacobian ^nJ with respect to the coordinate frame of link n is selected, the direct problem solution is obtained by using the backward recursive symbolic relations, derived in Subsections 2.2.3 and 2.2.5 (for parallel revolute joints). Afterwards, the expert program instructions, to evaluate the Jacobian elements, are written into the output file, using the symbolic recursive expressions derived in Subsections 2.3.2 and 2.3.3.

If the Jacobian ^OJ with respect to the base reference system is desired, the program lines which evaluate the left three columns of ^OT_i matrices, are generated according to the symbolic relations derived in Subsections 2.2.4 and 2.2.5. The evaluation of the position vector will be performed according to the backward symbolic relations developed in Subsections 2.2.3 and 2.2.5. The program instructions for evaluating the Jacobian ^OJ are formed using the equations derived in Subsections 2.4.2 and 2.4.3.

The expert program instructions are formed in the same manner as it was described in the examples given in the preceding sections. There are no multiplications by zero parameters or zero variables, nor are there any additions. Multiplication by ±1 is also avoided. These principles reduce the computational complexity of the symbolic kinematic model of a given manipulator. However, the model is valid only for the class of manipulators with the specific kinematic configuration. This means that the nonzero parameters, such as a_i, d_i, can be varied, and the same expert program used; however, the parameters with values equal to zero or ±1 (such as sines and cosines of the twist angles) no longer figure symbolically in the output program, and cannot be varied. If they are to be changed a new expert program should be generated by running the program package again with the modified kinematic parameters.

The program can be made in any suitable high level programming language. It contains only simple programming instructions: evaluation of sines and cosines, multiplication of real variables and additions. There are no program loops, no indexing of the variables, etc., so that upon compilation, execution will be most efficient.

2.5.2. Examples

In order to illustrate the process of automatic computer-aided generation of the kinematic model, we will here give some examples of expert
programs. These examples represent FORTRAN programs for evaluating the
kinematic model of the UMS-3B manipulator.

The first example is the program which calculates the manipulator hand
position, the orientation matrix with respect to the base reference
frame, and the Jacobian matrix with respect to the coordinate system
attached to the last link.

```
C
      SUBROUTINE EXPERT
C
C           MANIPULATOR UMS3B
C
      INCLUDE 'KIN6.COM'
C
      S1=SIN(Q1)
      C1=COS(Q1)
      S2=SIN(Q2)
      C2=COS(Q2)
      S4=SIN(Q4)
      C4=COS(Q4)
      S5=SIN(Q5)
      C5=COS(Q5)
      S6=SIN(Q6)
      C6=COS(Q6)
C
C   DIRECT KINEMATIC PROBLEM
C
      T6514=A6*C6
      T6524=A6*S6
      TP65= T6514+ A5
      T6411= C5*C6
      T6421= S5*C6
      T6412=-C5*S6
      T6422=-S5*S6
      T6414= C5*TP65
      T6424= S5*TP65
      T6311= C4*T6411+S4*S6
      T6321= S4*T6411-C4*S6
      T6312= C4*T6412+S4*C6
      T6322= S4*T6412-C4*C6
      T6313= C4*S5
      T6323= S4*S5
      T6314= C4*T6414+S4*T6524
      T6324= S4*T6414-C4*T6524
      T6234= T6424+Q3
      TP62= T6314+ A2
      T6111= C2*T6311-S2*T6421
      T6121= S2*T6311+C2*T6421
      T6112= C2*T6312-S2*T6422
      T6122= S2*T6312+C2*T6422
      T6113= C2*T6313+S2*C5
      T6123= S2*T6313-C2*C5
```

```
      T6114= C2*TP62-S2*T6234
      T6124= S2*TP62+C2*T6234
      T6011= C1*T6111-S1*T6321
      T6021= S1*T6111+C1*T6321
      T6031= T6121
      T6012= C1*T6112-S1*T6322
      T6022= S1*T6112+C1*T6322
      T6032= T6122
      T6013= C1*T6113-S1*T6323
      T6023= S1*T6113+C1*T6323
      T6033= T6123
      T6014= C1*T6114-S1*T6324
      T6024= S1*T6114+C1*T6324
      T6034= T6124
C
C    JACOBIAN WRT TO MAN. HAND
C
      J16= 0.
      J26= 0.
      J36= 0.
      J46= 0.
      J56= 0.
      J66= 1.
      J15= 0.
      J25= 0.
      J35=-TP65
      J45= S6
      J55= C6
      J65= 0.
      J14= T6411*T6524-T6414*S6
      J24= T6412*T6524-T6414*C6
      J34= S5*T6524
      J44= T6421
      J54= T6422
      J64=-C5
      J13= T6421
      J23= T6422
      J33=-C5
      J43= 0.
      J53= 0.
      J63= 0.
      J12=-T6311*T6234+TP62*T6421
      J22=-T6312*T6234+TP62*T6422
      J32=-T6313*T6234-TP62*C5
      J42=-T6321
      J52=-T6322
      J62=-T6323
      J11=-T6111*T6324+T6114*T6321
      J21=-T6112*T6324+T6114*T6322
      J31=-T6113*T6324+T6114*T6323
      J41= T6121
      J51= T6122
      J61= T6123
      RETURN
      END
```

The similarity between this program and the Equations (2.2.49)-(2.2.55) and (2.3.20)-(2.3.22), derived for the UMS-3B manipulator manually, according to the developed symbolic recursive relations, is evident. We see, that the program contains only multiplications, additions of real variables and evaluation of sines and cosines of joint angles. The equations which correspond to simple variables and are not output variables are omitted from the program.

The second example is the evaluation of the manipulator hand position, the orientation matrix and the Jacobian with respect to the base reference coordinate frame for the UMS-3B manipulator.

```
C
          SUBROUTINE EXPERT
C
          INCLUDE 'KIN.COM'
C
          S1=SIN(Q1)
          C1=COS(Q1)
          S2=SIN(Q2)
          C2=COS(Q2)
          S4=SIN(Q4)
          C4=COS(Q4)
          S5=SIN(Q5)
          C5=COS(Q5)
          S6=SIN(Q6)
          C6=COS(Q6)
C
C         DIRECT KINEMATIC PROBLEM
C
          T6514=A6*C6
          T6524=A6*S6
          TS101= S2*C1
          TS102= S2*S1
          T2011= C2*C1
          T2021= C2*S1
          TS301= S4*T2011+C4*S1
          TS302= S4*T2021-C4*C1
          TS303= S4*S2
          T4011= C4*T2011-S4*S1
          T4021= C4*T2021+S4*C1
          T4031= C4*S2
          TS401= S5*T4011+C5*TS101
          TS402= S5*T4021+C5*TS102
          TS403= S5*T4031-C5*C2
          T5011= C5*T4011-S5*TS101
          T5021= C5*T4021-S5*TS102
          T5031= C5*T4031+S5*C2
          TS501= S6*T5011-C6*TS301
          TS502= S6*T5021-C6*TS302
          TS503= S6*T5031-C6*TS303
          T6011= C6*T5011+S6*TS301
          T6021= C6*T5021+S6*TS302
          T6031= C6*T5031+S6*TS303
          T6012=-TS501
          T6022=-TS502
          T6032=-TS503
```

```
      T6013= TS401
      T6023= TS402
      T6033= TS403
C
C     VECTORS BETWEEN JOINTS AND MANIPULATOR TIP
C
      TP65= T6514+ A5
      T6414= C5*TP65
      T6424= S5*TP65
      T6314= C4*T6414+S4*T6524
      T6324= S4*T6414-C4*T6524
      T6234= T6424+Q3
      TP62= T6314+ A2
      T6114= C2*TP62-S2*T6234
      T6124= S2*TP62+C2*T6234
      T6014= C1*T6114-S1*T6324
      T6024= S1*T6114+C1*T6324
      T6034= T6124
C
C     JACOBIAN WITH RESPECT TO MAN. BASE
C
      J11=-T6024
      J21= T6014
      J31= 0.
      J41= 0.
      J51= 0.
      J61= 1.
      J12=-C1*T6124
      J22=-S1*T6124
      J32= T6114
      J42= S1
      J52=-C1
      J62= 0.
      J13=-TS101
      J23=-TS102
      J33= C2
      J43= 0.
      J53= 0.
      J63= 0.
      J14=-T2011*T6324-S1*T6314
      J24=-T2021*T6324+C1*T6314
      J34=-S2*T6324
      J44=-TS101
      J54=-TS102
      J64= C2
      J15=-T4011*T6424-TS101*T6414
      J25=-T4021*T6424-TS102*T6414
      J35=-T4031*T6424+C2*T6414
      J45= TS301
      J55= TS302
      J65= TS303
      J16=-T5011*T6524+TS301*T6514
      J26=-T5021*T6524+TS302*T6514
      J36=-T5031*T6524+TS403*T6514
      J46= TS401
      J56= TS402
      J66= TS403
      RETURN
      END
>
```

2.5.3. Numerical aspects

In this subsection we will analyze computational complexity of the kinematic models of several manipulator configurations, obtained by the developed algorithm. We will also compare the computational complexity of the symbolic models with the numeric models.

Table T.2.2 indicates the number of floating-point multiplications and additions required for evaluating the kinematic model of several manipulator configurations. The model includes the evaluation of the hand position, the orientation matrix and the Jacobian ^nJ with respect to the coordinate system of the last link. The execution times for the corresponding expert programs, measured on minicomputer PDP 11/70 and on a microcomputer based on microprocessor INTEL 8086 with arithmetic coprocessor 8087, are also indicated in the table. It should be observed that all revolute joints of the manipulators considered have nonzero lengths $(a_i \neq 0)$.

We can see from this table that the total number of multiplications and additions does not exceed 83 and 46, respectively, for the manipulator with 6 revolute nonparallel joints where all the links are of nonzero lengths. Most industrial manipulators, however, are of less complicated kinematic structure. The number of operations carried out and the times of execution are very small. A comparison of the computational complexity between this symbolic model and the corresponding numeric model can not be considered to be fair. Nevertheless, we will here point out that the numeric complexity of the purely numeric evaluation of the model would be several times higher than the numeric complexity obtained, even in the case when the most efficient method of computing the Jacobian is considered [14].

The algorithm developed in this chapter provides for minimal computational complexity in the kinematic model. This fact has not been strictly proved. Examples, however indicate that further reduction of the number of operations is almost impossible, and where it is possible the reduction would be practically negligible (see the discussion following Equation (2.3.17)).

As we see, the program package generates equations for evaluating the homogeneous transformation ^oT_n between system n and the base reference system, and either matrix ^nJ or ^oJ. When the evaluation of only one of these matrices is required, the corresponding expert program may be

Manipulator	Multiplications	Additions	sin/cos	Execution time [ms] PDP11/70	time [ms] INTEL 8086/8087
	2	0	2	0.32	1.9
	27	10	6	0.9	5.9
	49	22	8	1.32	9.2
	65	27	12	1.5	10.5
	71	33	10	1.7	11.2
	83	46	12	2.17	13.8

Table T.2.2. Computational complexity of the kinematic models of various manipulator configurations

obtained by simple rejecting the lines which are not necessary for the particular evaluation.

During the process of generating the symbolic kinematic model we have considered only the evaluation of the Jacobian matrices which correspond to the manipulator end point, since these are most often required in motion synthesis. However, the Jacobian which corresponds to some other point on the arm may also be required (e.g. see the algorithm for obstacle avoidance presented in Section 6.5). The software package which has been developed can also be used for generating the equations for evaluating the Jacobian for any arbitrary point on the arm. In that case the recursive relation derived would be applied in the narrower range $i=1,...,k$ instead of $i=1,...,n$, where k denotes the serial number of the link to which the point belongs.

Besides, the method for deriving the recursive symbolic relations for evaluating various kinematic variables which has been presented in this chapter, may also be used in the evaluation of any other kinematic variable, such as vector cross products $\vec{e}_j \times \vec{r}_{jk}$ and the like, which are necessary in the dynamic model construction.

We may also note that in the kinematic symbolic model construction terms of the type $\sin^2 q$, $\cos^2 q$ did not appear, or if they did appear they could always be eliminated. Thus the kinematic symbolic model depends only on the joint coordinates themselves (for sliding joints) and on the sines and cosines of joint angles (for revolute joints). It should be pointed out that this would not apply to the dynamic manipulator model, as was strictly proved in [16].

2.6. Summary

In this chapter a method has been presented for the automatic, computer-aided generation of a computer program for evaluating the symbolic kinematic model of an arbitrary serial-link manipulator. The computer program obtained contains the minimal number of floating-point multiplications and additions. It can be made in any suitable programming language and it is ready for compilation.

This approach to kinematic modelling of manipulation robots is preferred mainly because of its automation, which results in the elimination of the errors which could easily occur in the manual derivation of these equations. With regard to the advantages of this automatization, the necessity for numeric evaluation of kinematic models is completely eliminated in the motion synthesis of industrial robots. The numeric evaluation of the model remains significant in general software packages for simulating robot performance, which should not depend on robot configuration.

Appendix I
Direct Kinematic Problem for the Arthropoid Manipulator

In order to illustrate the application of the backward and forward recursive symbolic relations, derived in Subsection 2.2.5 specially for
revolute joints with parallel joint axes, we will here solve the direct
kinematic problem for the arthropoid manipulator shown in Fig. A.1.
This example will also help us to get more insight into the order of
the savings accomplished by taking the parallelism into account analytically.

The arthropoid manipulator has 6 revolute joints. The axes of joints 2,
3 and 4 are horizontal and always parallel. The Denavit-Hartenberg's
kinematic parameters of the manipulator are shown in Table T.1.

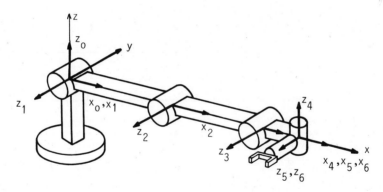

Fig. A.1. The arthropoid manipulator

Link	Joint coordinate	α_i	a_i	d_i	$\cos\alpha_i$	$\sin\alpha_i$
1	q_1	90°	0	0	0	1
2	q_2	0°	a_2	0	1	0
3	q_3	0°	a_3	0	1	0
4	q_4	-90°	a_4	0	0	-1
5	q_5	90°	0	0	0	1
6	q_6	0°	0	0	1	0

Table T.1. Kinematic parameters for the arthropoid manipulator

The homogeneous transformations $^{i-1}T_i$, $i=1,\ldots,6$ for this manipulator are given by

$$^0T_1 = \begin{bmatrix} C_1 & 0 & S_1 & 0 \\ S_1 & 0 & -C_1 & 0 \\ 0 & 1 & 0 & 0 \\ 0 & 0 & 0 & 1 \end{bmatrix}$$

(A.1.1)

$$^1T_2 = \begin{bmatrix} C_2 & -S_2 & 0 & a_2C_2 \\ S_2 & C_2 & 0 & a_2S_2 \\ 0 & 0 & 1 & 0 \\ 0 & 0 & 0 & 1 \end{bmatrix}$$

(A.1.2)

$$^2T_3 = \begin{bmatrix} C_3 & -S_3 & 0 & a_3C_3 \\ S_3 & C_3 & 0 & a_3S_3 \\ 0 & 0 & 1 & 0 \\ 0 & 0 & 0 & 1 \end{bmatrix}$$

(A.1.3)

$$^3T_4 = \begin{bmatrix} C_4 & 0 & -S_4 & a_4C_4 \\ S_4 & 0 & C_4 & a_4S_4 \\ 0 & -1 & 0 & 0 \\ 0 & 0 & 0 & 1 \end{bmatrix}$$

(A.1.4)

$$^4T_5 = \begin{bmatrix} C_5 & 0 & S_5 & 0 \\ S_5 & 0 & -C_5 & 0 \\ 0 & 1 & 0 & 0 \\ 0 & 0 & 0 & 1 \end{bmatrix}$$

(A.1.5)

$$^5T_6 = \begin{bmatrix} C_6 & -S_6 & 0 & 0 \\ S_6 & C_6 & 0 & 0 \\ 0 & 0 & 1 & 0 \\ 0 & 0 & 0 & 1 \end{bmatrix}$$

(A.1.6)

Let us first consider the direct problem solution according to backward recursive relations. Premultiplying the above matrices we obtain

$$
^5T_6 = \begin{bmatrix} C_6 & -S_6 & 0 & 0 \\ S_6 & C_6 & 0 & 0 \\ 0 & 0 & 1 & 0 \\ 0 & 0 & 0 & 1 \end{bmatrix}
\tag{A.1.7}
$$

$$
^4T_6 = \begin{bmatrix} C_5C_6 & -C_5S_6 & S_5 & 0 \\ S_5C_6 & -S_5S_6 & -C_5 & 0 \\ S_6 & C_6 & 0 & 0 \\ 0 & 0 & 0 & 1 \end{bmatrix}
\tag{A.1.8}
$$

$$
^3T_6 = \begin{bmatrix} C_4C_5C_6-S_4S_6 & -C_4C_5S_6-S_4C_6 & C_4S_5 & a_4C_4 \\ S_4C_5C_6+C_4S_6 & -S_4C_5S_6+C_4C_6 & S_4S_5 & a_4S_4 \\ -S_5C_6 & S_5S_6 & C_5 & 0 \\ 0 & 0 & 0 & 1 \end{bmatrix}
\tag{A.1.9}
$$

$$
^2T_6 = \begin{bmatrix} C_{3,4}C_5C_6-S_{3,4}S_6 & -C_{3,4}C_5S_6-S_{3,4}C_6 & C_{3,4}S_5 & a_4C_{3,4}+a_3C_3 \\ S_{3,4}C_5C_6+C_{3,4}S_6 & -S_{3,4}C_5S_6+C_{3,4}C_6 & S_{3,4}S_5 & a_4S_{3,4}+a_3S_3 \\ -S_5C_6 & S_5S_6 & C_5 & 0 \\ 0 & 0 & 0 & 1 \end{bmatrix}
\tag{A.1.10}
$$

$$
^1T_6 = \begin{bmatrix} C_{2,4}C_5C_6-S_{2,4}S_6 & -C_{2,4}C_5C_6-S_{2,4}C_6 & C_{2,4}S_5 & a_4C_{2,4}+a_3C_{2,3}+a_2C_2 \\ S_{2,4}C_5C_6+C_{2,4}S_6 & -S_{2,4}C_5C_6+C_{2,4}C_6 & S_{2,4}S_5 & a_4S_{2,4}+a_3S_{2,3}+a_2S_2 \\ -S_5C_6 & S_5S_6 & C_5 & 0 \\ 0 & 0 & 0 & 1 \end{bmatrix}
\tag{A.1.11}
$$

$$
^0T_6 = \begin{bmatrix} T6011 & T6012 & T6013 & T6014 \\ T6021 & T6022 & T6023 & T6024 \\ T6031 & T6032 & T6033 & T6034 \\ 0 & 0 & 0 & 1 \end{bmatrix}
$$

where

$$T6011 = C_1(C_{2,4}C_5C_6 - S_{2,4}S_6) - S_1S_5C_6$$

$$T6021 = S_1(C_{2,4}C_5C_6 - S_{2,4}S_6) + C_1S_5C_6$$

$$T6031 = S_{2,4}C_5C_6 + C_{2,4}S_6$$

$$T6012 = -C_1(C_{2,4}C_5S_6 + S_{2,4}C_6) + S_1S_5S_6$$

$$T6022 = -S_1(C_{2,4}C_5S_6 + S_{2,4}C_6) - C_1S_5S_6$$

$$T6032 = -S_{2,4}C_5S_6 + C_{2,4}C_6 \qquad\qquad (A.1.12)$$

$$T6013 = C_1C_{2,4}S_5 + S_1C_5$$

$$T6023 = S_1C_{2,4}S_5 - C_1C_5$$

$$T6033 = S_{2,4}S_5$$

$$T6014 = C_1(a_4C_{2,4} + a_3C_{2,3} + a_2C_2)$$

$$T6024 = S_1(a_4C_{2,4} + a_3C_{2,3} + a_2C_2)$$

$$T6034 = a_4S_{2,4} + a_3S_{2,3} + a_2S_2$$

Let us now write down the set of equations by which the above variables would be evaluated optimally, with the minimal number of multiplications and additions. For this we will make use of the backward recursive symbolic relations derived either for ordinary revolute joints or for parallel revolute joints.

For this manipulator the first parallel axis is z_1 and the last z_3, yielding k=2, K=4. The transformation matrices $^{i-1}T_n$, i=n, n-1,...K+1, i.e. $^5T_6$ and $^4T_6$ are obtained from the standard recursive symbolic relations for revolute joints (2.2.37) - (2.2.44). Thus we obtain the following set of equations

$$TR661 = 0$$

$$TR662 = 1$$

$$TR663 = 0$$

$$TR664 = 0$$

$$TP66 = 0$$

$$T6511 = C_6$$

$$T6521 = S_6$$

$$T6531 = 0$$

$$T6512 = -S_6 \qquad\qquad (A.1.13)$$

$$T6522 = C_6$$

$$T6532 = 0$$

$$T6513 = 0$$

$$T6523 = 0$$

$$T6533 = 1$$

$$T6514 = 0$$

$$T6524 = 0$$

$$T6534 = 0$$

and

$$TR651 = S_6$$

$$TR652 = C_6$$

$$TR653 = 0$$

$$TR654 = 0$$

$$TP65 = 0$$

$$T6411 = C_5 * C_6$$

$$T6421 = S_5 * C_6$$

$$T6431 = S_6$$

$$T6412 = -C_5 * S_6 \hspace{4cm} (A.1.14)$$

$$T6422 = -S_5 * S_6$$

$$T6432 = C_6$$

$$T6413 = S_5$$

$$T6423 = -C_5$$

$$T6433 = 0$$

$$T6414 = 0$$

$$T6424 = 0$$

$$T6434 = 0$$

The next two transformations need not be evaluated, since they represent the medial transformations $^{K-1}T_n = {}^3T_6$ and $^{K-2}T_n = {}^kT_n = {}^2T_6$ corresponding to the parallel joints. The values of the elements of these two matrices do not figure in the expressions for the elements of $^{k-1}T_n = {}^1T_6$. We can, therefore, write down directly the set of equations for the matrix $^1T_6$, according to (2.2.83) - (2.2.87) and (A.1.14)

$TR641 = S_6$

$TR642 = C_6$

$TR643 = 0$

$TR644 = 0$

$TP64 = a_4$

$T6111 = C_{2,4}*T6411-S_{2,4}*S_6$

$T6121 = S_{2,4}*T6411+C_{2,4}*S_6$

$T6131 = -T6421$

$T6112 = C_{2,4}*T6412-S_{2,4}*C_6$

$T6122 = S_{2,4}*T6412+C_{2,4}*C_6$

$T6132 = -T6422$ $\qquad\qquad$ (A.1.15)

$T6113 = C_{2,4}*S_5$

$T6123 = S_{2,4}*S_5$

$T6133 = C_5$

$T61141 = C_{2,4}*a_4$

$T61142 = T61141+a_3*C_{2,3}$

$T6114 = T61142+a_2*C_2$

$T61241 = S_{2,4}*a_4$

$T61242 = T61241+a_3*S_{2,3}$

$T6124 = T61242+a_2*S_2$

$T6134 = 0$

The elements of $^OT_6$ matrix are obtained from the standard recursive symbolic relations (2.2.37) - (2.2.44) and (A.1.15)

$TR611 = T6421$

$TR612 = T6422$

$TR613 = -C_5$

$TR614 = 0$

$TP61 = T6114$

$T6011 = C_1*T6111-S_1*T6421$

$T6021 = S_1*T6111+C_1*T6421$

$T6031 = T6121$

$$T6012 = C_1*T6112-S_1*T6422$$

$$T6022 = S_1*T6112+C_1*T6422 \qquad\qquad (A.1.16)$$

$$T6032 = T6122$$

$$T6013 = C_1*T6113+S_1*C_5$$

$$T6023 = S_1*T6113-C_1*C_5$$

$$T6033 = T6123$$

$$T6014 = C_1*T6114$$

$$T6024 = S_1*T6114$$

$$T6034 = T6124$$

The evaluation of matrix $^0T_6$ according to the above set of equations, requires 42 multiplications and 18 additions and 12 transedental function calls (the sines and cosines of the sums of angles take 8 multiplications and 4 additions). If the problem was solved without any care of the parallel joints, it would require 62 multiplications and 27 additions and 12 transcedental function calls. Thus, the savings are 20 multiplications and 9 additions, or about 30%.

One should note that the number of saved numerical operations depends on the number of joints with parallel axes. It increases with the number of parallel joints, since the number of medial transformations which are not evaluated, also increases.

Let us now illustrate the application of the forward recursive relations in optimal evaluation of the transformation ^0T_n for the arthropoid manipulator in Fig. A.1.

By postmultiplying the homogeneous transformation matrices $^{i-1}T_i$, $i=1,\ldots,n$, given by (A.1.1) - (A.1.6), and compressing the expressions we obtain the analytical forms of matrices ^0T_i, $i=1,\ldots,n$

$$^0T_1 = \begin{bmatrix} C_1 & 0 & S_1 & 0 \\ S_1 & 0 & -C_1 & 0 \\ 0 & 1 & 0 & 0 \\ 0 & 0 & 0 & 1 \end{bmatrix} \qquad\qquad (A.1.17)$$

$$
{}^{0}T_2 = \begin{bmatrix}
C_2C_1 & -S_2C_1 & S_1 & a_2C_2C_1 \\
C_2S_1 & -S_2S_1 & -C_1 & a_2C_2S_1 \\
S_2 & C_2 & 0 & a_2S_2 \\
0 & 0 & 0 & 1
\end{bmatrix}
\tag{A.1.18}
$$

$$
{}^{0}T_3 = \begin{bmatrix}
C_{2,3}C_1 & -S_{2,3}C_1 & S_1 & (a_3C_{2,3}+a_2C_2)C_1 \\
C_{2,3}S_1 & -S_{2,3}S_1 & -C_1 & (a_3C_{2,3}+a_2C_2)S_1 \\
S_{2,3} & C_{2,3} & 0 & a_3S_{2,3}+a_2S_2 \\
0 & 0 & 0 & 1
\end{bmatrix}
\tag{A.1.19}
$$

$$
{}^{0}T_4 = \begin{bmatrix}
C_{2,4}C_1 & -S_1 & -S_{2,4}C_1 & (a_4C_{2,4}+a_3C_{2,3}+a_2C_2)C_1 \\
C_{2,4}S_1 & C_1 & -S_{2,4}S_1 & (a_4C_{2,4}-a_3C_{2,3}+a_2C_2)S_1 \\
S_{2,4} & 0 & C_{2,4} & a_4S_{2,4}+a_3S_{2,3}+a_2S_2 \\
0 & 0 & 0 & 1
\end{bmatrix}
\tag{A.1.20}
$$

$$
{}^{0}T_5 = \begin{bmatrix}
C_5C_{2,4}C_1-S_5S_1 & -S_{2,4}C_1 & S_5C_{2,4}C_1+C_5S_1 & (a_4C_{2,4}+a_3C_{2,3}+a_2C_2)C_1 \\
C_5C_{2,4}S_1+S_5C_1 & -S_{2,4}S_1 & S_5C_{2,4}S_1-C_5C_1 & (a_4C_{2,4}+a_3C_{2,3}+a_2C_2)S_1 \\
C_5S_{2,4} & C_{2,4} & S_5S_{2,4} & a_4S_{2,4}+a_3S_{2,3}+a_2S_2 \\
0 & 0 & 0 & 1
\end{bmatrix}
\tag{A.1.21}
$$

$$
{}^{0}T_6 = \begin{bmatrix}
T6011 & T6012 & T6013 & T6014 \\
T6021 & T6022 & T6023 & T6024 \\
T6031 & T6032 & T6033 & T6034 \\
0 & 0 & 0 & 1
\end{bmatrix}
$$

where

$$
\begin{aligned}
T6011 &= C_6(C_5C_{2,4}C_1-S_5S_1)-S_6S_{2,4}C_1 \\
T6021 &= C_6(C_5C_{2,4}S_1+S_5C_1)-S_6S_{2,4}S_1 \\
T6031 &= C_6C_5S_{2,4}+S_6C_{2,4} \\
T6012 &= -S_6(C_5C_{2,4}C_1-S_5S_1-C_6S_{2,4}C_1 \\
T6022 &= -S_6(C_5C_{2,4}S_1+S_5C_1)-C_6S_{2,4}S_1 \\
T6032 &= -S_6C_5S_{2,4}+C_6C_{2,4} \\
T6013 &= S_5C_{2,4}C_1+C_5S_1
\end{aligned}
\tag{A.1.22}
$$

$$T6023 = S_5C_{2,4}S_1 - C_5C_1$$

$$T6033 = S_5S_{2,4}$$

$$T6014 = (a_4C_{2,4} + a_3C_{2,3} + a_2C_2)C_1$$

$$T6024 = (a_4C_{2,4} + a_3C_{2,3} + a_2C_2)S_1$$

$$T6034 = a_4S_{2,4} + a_3S_{2,3} + a_2S_2$$

Let us now write down the set of equations which provide for the most efficient evaluation of the transformation $^{O}T_6$, using the forward recursive symbolic relations derived either for ordinary revolute joints or for parallel revolute joints. The fourth columns of matrices $^{O}T_i$, $i=1,\ldots,n$ will not be evaluated. Instead, the manipulator hand position will be determined by backward recursive relations.

The first joint is an ordinary revolute joint, so that matrix $^{O}T_1 = {}^{O}T_{k-1}$ is obtained from Equations (2.2.61) - (2.2.64) for $i=1$:

$$TS001 = S_1$$

$$TS002 = C_1$$

$$TS003 = 0$$

$$T1011 = C_1$$

$$T1021 = S_1$$

$$T1031 = 0$$

$$T1012 = 0$$ (A.1.23)

$$T1022 = 0$$

$$T1032 = 1$$

$$T1013 = S_1$$

$$T1023 = -C_1$$

$$T1033 = 0$$

For this manipulator $k=2$ and $K=4$, as we have already seen (see Fig. A.1). The intermediate transformations $^{O}T_k = {}^{O}T_2$ and $^{O}T_{k+1} = {}^{O}T_{K-1} = {}^{O}T_3$ need not be evaluated. Equations (2.2.96) - (2.2.97) yield the elements of the transformation $^{O}T_K = {}^{O}T_4$ directly

$$TS241 = S_{2,4}*C_1$$

$$TS242 = S_{2,4}*S_1$$

$$TS243 = -C_{2,4}$$

$$T4011 = C_{2,4}*C_1$$

$$T4021 = C_{2,4}*S_1$$

$$T4031 = S_{2,4}$$

$$T4012 = -S_1$$

$$T4022 = C_1$$

$$T4032 = 0$$

$$T4013 = -TS241$$

$$T4023 = -TS242$$

$$T4033 = C_{2,4}$$

(A.1.24)

The left three columns of matrices $^{O}T_5$ and $^{O}T_6$ are again obtained from Equations (2.2.61) - (2.2.64)

$$TS401 = S_5*T4011+C_5*S_1$$

$$TS402 = S_5*T4021-C_5*C_1$$

$$TS403 = S_5*S_{2,4}$$

$$T5011 = C_5*T4011-S_5*S_1$$

$$T5021 = C_5*T4021+S_5*C_1$$

$$T5031 = C_5*S_{2,4}$$

$$T5012 = -TS241$$

$$T5022 = -TS242$$

$$T5032 = C_{2,4}$$

$$T5013 = TS401$$

$$T5023 = TS402$$

$$T5033 = TS403$$

(A.1.25)

$$TS501 = S_6*T5011+C_6*TS241$$

$$TS502 = S_6*T5021+C_6*TS242$$

$$TS503 = S_6*T5031-C_6*C_{2,4}$$

$$T6011 = C_6*T5011-S_6*TS241$$

$$T6021 = C_6*T5021-S_6*TS242$$

$$T6031 = C_6*T5031+S_6*C_{2,4}$$

$$T6012 = -TS501$$

$$T6022 = -TS502 \qquad\qquad (A.1.26)$$

$$T6032 = -TS503$$

$$T6013 = T5013$$

$$T6023 = T5023$$

$$T6033 = T5033$$

The evaluation of the fourth columns of matrices ${}^iT_6$, $i=6,5,\ldots,1$ for the arthropoid manipulator has already been analized in the Equations (A.1.13) - (A.1.16). They are obtained by applying the symbolic relations for ordinary revolute joints ((2.2.40), (2.2.42), (2.2.44)) and by utilizing the expressions (2.2.86) and (2.2.87). For the sake of clarity we will here repeat this set of equations.

The position vectors in matrices ${}^5T_6$, ${}^4T_6$ are obtained from standard recursive expressions

$$TR664 = 0$$

$$TP66 = 0$$

$$T6514 = 0 \qquad\qquad (A.1.27)$$

$$T6524 = 0$$

$$T6534 = 0$$

$$TR654 = 0$$

$$TP65 = 0$$

$$T6414 = 0 \qquad\qquad (A.1.28)$$

$$T6424 = 0$$

$$T6434 = 0$$

The fourth columns of the intermediate matrices ${}^3T_6$ and ${}^2T_6$ need not be computed. We evaluate the fourth column of ${}^1T_6 = {}^{k-1}T_n$ directly, according to (2.2.86), (2.2.87)

TR644 = 0

TP64 = a_4

T61141 = $C_{2,4}*a_4$

T61142 = T61141+$C_{2,3}*a_3$

T6114 = T61142*C_2*a_2 $\hspace{3cm}$ (A.1.29)

T61241 = $S_{2,4}*a_4$

T61242 = T61241+$S_{2,3}*a_3$

T6124 = T61242*S_2*a_2

T6134 = 0

The manipulator hand position with respect to the base reference frame is obtained from

TR614 = 0

TP61 = T6114

T6014 = C_1*T6114 $\hspace{3cm}$ (A.1.30)

T6024 = S_1*T6114

T6034 = T6124

The total number of operations required is 42 multiplications and 18 additions and 12 transcedental function calls. If the problem was solved by forward recursive relations, but without taking into account the parallelism between the joint axes, it would require about 63 multiplications 29 additions and 12 transedental function calls.

Appendix II

The Jacobian with Respect to the Hande Coordinate Frame for the Arthropoid Manipulator

In order to estimate the computational savings achieved by introducing the special evaluation of the Jacobian columns, which correspond to revolute joints with parallel joint axes, we will here derive the set of equations for computing the Jacobian with respect to the hand coordinate frame for the arthropoid manipulator presented in Fig. A.1 in Appendix I.

We will first derive analytical expressions for the elements of matrix ^nJ. Applying the general Equations (2.3.2) and (2.3.3) to the homogeneous transformation matrices derived for this manipulator in Appendix I (Eqs. (A.1.7) – (A.1.12)), we obtain

$$J_{11} = -(C_{2,4}a_4 + C_{2,3}a_3 + C_2 a_2) S_5 C_6$$

$$J_{21} = -(C_{2,4}a_4 + C_{2,3}a_3 + C_2 a_2) S_5 S_6$$

$$J_{31} = -(C_{2,4}a_4 + C_{2,3}a_3 + C_2 a_2) C_5$$

$$J_{41} = S_{2,4} C_5 C_6 + C_{2,4} S_6$$

$$J_{51} = -S_{2,4} C_5 S_6 + C_{2,4} C_6$$

$$J_{61} = S_{2,4} S_5$$

$$(A.2.1)$$

$$J_{12} = S_6 (a_4 + a_3 C_4 + a_2 C_{3,4}) + C_5 C_6 (a_3 S_4 + a_2 S_{3,4})$$

$$J_{22} = C_6 (a_4 + a_3 C_4 + a_2 C_{3,4}) + S_5 C_6 (a_3 S_4 + a_2 S_{3,4})$$

$$J_{32} = S_6 (a_3 S_4 + a_2 S_{3,4})$$

$$(A.2.2)$$

$$J_{42} = -S_5 C_6$$

$$J_{52} = S_5 S_6$$
$$J_{62} = C_5$$

$$J_{13} = S_6 (a_4 + a_3 C_4) + C_5 C_6 a_3 S_4$$
$$J_{23} = C_6 (a_4 + a_3 C_4) + S_5 C_6 a_3 S_4$$
$$J_{33} = S_6 a_3 S_4$$
$$J_{43} = -S_5 C_6$$
$$J_{53} = S_5 S_6$$
$$J_{63} = C_5$$

$$\text{(A.2.3)}$$

$$J_{14} = a_4 S_6$$
$$J_{24} = a_4 C_6$$
$$J_{34} = 0$$
$$J_{44} = -S_5 C_6$$
$$J_{54} = S_5 S_6$$
$$J_{64} = C_5$$

$$\text{(A.2.4)}$$

$$J_{15} = 0$$
$$J_{25} = 0$$
$$J_{35} = 0$$
$$J_{45} = S_6$$
$$J_{55} = C_6$$
$$J_{65} = 0$$

$$\text{(A.2.5)}$$

$$J_{16} = 0$$
$$J_{26} = 0$$
$$J_{36} = 0$$
$$J_{46} = 0$$
$$J_{56} = 0$$
$$J_{66} = 1$$

$$\text{(A.2.6)}$$

We will now write down the corresponding set of equations which provide for optimal evaluation of the above expressions. We will apply the method derived in Subsection 2.3.3, and use the set of equations (A.1.13) - (A.1.16) in computing matrix $^{0}T_{n}$ for the arthropoid manipulator. Since the first joint is an ordinary revolute joint, the first column is obtained by applying relations (2.3.17) - (2.3.18).

$$J11 = -T6114*T6421$$
$$J21 = -T6114*T6422$$
$$J31 = -T6114*C_5$$
$$J41 = T6121 \hspace{2cm} (A.2.7)$$
$$J51 = T6122$$
$$J61 = T6123$$

In this manipulator the axes of joints 2, 3 and 4 are parallel, so that $k=2$ and $K=4$. The columns $^{n}j^{(i)}$ for $i=k=2$, $i=k+1=K-1=3$ are evaluated according to the specially derived relations (2.3.25) - (2.3.28)

$$JP1 = a_4+a_3*C_4$$
$$JP2 = JP1+a_2*C_{3,4}$$
$$JR1 = a_3*S_4$$
$$JR2 = JR1+a_2*S_{3,4}$$
$$J12 = S_6*JP2+T6411*JR2$$
$$J22 = C_6*JP2+T6412*JR2 \hspace{2cm} (A.2.8)$$
$$J32 = S_6*JR2$$
$$J42 = -T6421$$
$$J52 = -T6422$$
$$J62 = C_5$$

and

$$J13 = S_6*JP1+T6411*JR1$$
$$J23 = C_6*JP1+T6412*JR1$$
$$J33 = S_6*JR1$$
$$J43 = -T6421 \hspace{2cm} (A.2.9)$$
$$J53 = -T6422$$
$$J63 = C_5$$

The columns for i=4,5,6 are again obtained from (2.3.17) - (2.3.18)

$$J14 = a_4*S_6$$
$$J24 = a_4*C_6$$
$$J34 = 0$$
$$J44 = -T6421$$
$$J54 = -T6422$$
$$J64 = C_5$$

(A.2.10)

$$J_{15} = 0$$
$$J_{25} = 0$$
$$J_{35} = 0$$
$$J_{45} = S_6$$
$$J_{55} = C_6$$
$$J_{65} = 0$$

(A.2.11)

$$J_{16} = 0$$
$$J_{26} = 0$$
$$J_{36} = 0$$
$$J_{46} = 0$$
$$J_{56} = 0$$
$$J_{66} = 1$$

(A.2.12)

We see that the evaluation of the Jacobian matrix with respect to the hand coordinate frame requires 23 multiplications and 9 additions. On the other hand if all the intermediate transformations for parallel joints were determined, the Jacobian matrix would be obtained by only 17 new multiplications and 6 additions. Disregarding this illusory loss, if the evaluation of both the matrices $^{O}T_n$ and ^{n}J is considered, the total savings are still 14 multiplies and 6 additions, with respect to the case where the parallel joints are treated as ordinary revolute joints.

Appendix III

The Jacobian with Respect to the Base Coordinate Frame for the Arthropoid Manipulator

In order to illustrate the evaluation of the Jacobian with respect to the base coordinate frame, for the manipulators which have revolute joints with parallel axes, we will here consider the arthropoid, 6 degree-of-freedom manipulator shown in Fig. A.1 in Appendix I. This example will help us to estimate the computational savings achieved by introducing the special evaluation of the Jacobian columns which correspond to parallel joints.

We will first derive the analytical expression for the Jacobian elements, using the Equations (2.4.10) - (2.4.11). The analytical expression for the elements of matrices ${}^{0}T_{i-1}$ for this manipulator have been derived in Appendix I, Equations (A.1.17) - (A.1.22), while the fourth columns of matrices ${}^{i-1}T_{n}$ are given by (A.1.7) - (A.1.12). We thus obtain

$$J_{11} = -(a_4 C_{2,4} + a_3 C_{2,3} + a_2 C_2) C_1$$

$$J_{21} = (a_4 C_{2,4} + a_3 C_{2,3} + a_2 C_2) S_1$$

$$J_{31} = 0$$

$$J_{41} = 0$$

$$J_{51} = 0$$

$$J_{61} = 1$$

$$J_{12} = -(a_4 S_{2,4} + a_3 S_{2,3} + a_2 S_2) C_1$$

$$J_{22} = -(a_4 S_{2,4} + a_3 S_{2,3} + a_2 S_2) S_1$$

$$J_{32} = a_4 C_{2,4} + a_3 C_{2,3} + a_2 C_2$$

$$J_{42} = S_1$$

$$J_{52} = -C_1$$

$$J_{62} = 0$$

$$J_{13} = -(S_{2,4}a_4 + S_{2,3}a_3)C_1$$

$$J_{23} = -(S_{2,4}a_4 + S_{2,3}a_3)S_1$$

$$J_{33} = C_{2,4}a_4 + C_{2,3}a_3$$

$$J_{43} = S_1$$

$$J_{53} = -C_1$$

$$J_{63} = 0$$

$$J_{14} = -S_{2,4}a_4C_1$$

$$J_{24} = -S_{2,4}a_4S_1$$

$$J_{34} = C_{2,4}a_4$$

$$J_{44} = S_1$$

$$J_{54} = -C_1$$

$$J_{64} = 0$$

$$J_{15} = 0$$

$$J_{25} = 0$$

$$J_{35} = 0$$

$$J_{45} = -S_{2,4}C_1$$

$$J_{55} = -S_{2,4}S_1$$

$$J_{65} = C_{2,4}$$

$$J_{16} = 0$$

$$J_{26} = 0$$

$$J_{36} = 0$$

$$J_{46} = S_5C_{2,4}C_1 + C_5S_1$$

$$J_{56} = S_5C_{2,4}S_1 - C_5C_1$$

$$J_{66} = S_5S_{2,4}$$

The corresponding set of equations to evaluate the above analytical expressions optimally, are obtained by applying Equations (2.4.10), (2.4.11) and specially derived Equations (2.4.18) - (2.4.19) are used for parallel revolute joints. The equations which evaluate the elements of homogeneous transformations of this manipulator, have been derived in Appendix I.

The first two columns $o_j{}^{(i)}$, i=1, i=k=2 are obtained from the equations for ordinary revolute joints (2.4.10) - (2.4.11)

     J11 = -T6024

     J21 = T6014

     J31 = 0

     J41 = 0

     J51 = 0

     J61 = 1

     J12 = $-C_1$*T6124

     J22 = $-S_1$*T6124

     J32 = T6114

     J42 = S_1

     J52 = $-C_1$

     J62 = 0

The columns $o_j{}^{(i)}$, i=k+1=3, i=k+2=K=4 are obtained from Equations (2.4.18) - (2.4.19)

     J13 = $-C_1$*T61242

     J23 = $-S_1$*T61242

     J33 = T61142

     J43 = S_1

     J53 = $-C_1$

     J63 = 0

     J14 = $-C_1$*T61241

     J24 = $-S_1$*T61241

     J34 = T61141

     J44 = S_1

     J54 = $-C_1$

     J64 = 0

The columns $o_j{}^{(i)}$, i=5, 6 are again obtained from (2.4.10), (2.4.11)

J15 = 0

J25 = 0

J35 = 0

J45 = T4013

J55 = T4023

J66 = T4033

J16 = 0

J26 = 0

J36 = 0

J46 = T5013

J56 = T5023

J66 = T5033

We recognize that the evaluation of the above equations requires only 6 additional multiplications and no additions upon the evaluation of $^{O}T_6$ matrix. If this Jacobian matrix with respect to the base coordinate frame was computed without special treatment of the parallel joints, it would require 14 multiplications and 6 additions. This computational saving together with the saving of 21 multiplications and 11 additions achieved in evaluating the matrix $^{O}T_6$, makes up the total saving of 29 multiplications and 17 additions. This is a considerable proportionate computational saving compared with the total number of numeric operations required for the evaluation of the Jacobian ^{O}J and the transformation $^{O}T_n$, which in this example amounts to 48 multiplications and 18 additions.

Chapter 3
Inverse Kinematic Problem

3.1. Introduction

In the previous chapters we have been concerned with the direct kinematic problem, i.e. the evaluation of end-effector position and orientation, as well as the matrix relating between the linear and angular hand velocities and joint rates.

In this chapter we will deal with the inverse kinematic problem. It involves the computation of the joint coordinates which correspond to a given hand position and orientation described by the external coordinates vector x_e, i.e. obtaining the inverse transformation

$$q = f^{-1}(x_e) \qquad\qquad (3.1.1)$$

where $q \in Q \subset R^n$, $x_e \in X_e \subset R^m$. In general, the solution to the inverse kinematic problem is not unique. For nonredundant manipulators, where the number of degrees of freedom is equal to the number of external coordinates $(n=m)$, several mappings may exist $f^{-1}: x_e \to q$, so that a finite set of solutions exists $\{q^j, j \in J\}$ in which $f(q^j) = x_e$, where J is a set of integer indices. With redundant manipulators, where $n > m$, there is an infinite set of solutions q which answer to $f(q) = x_e$. We will consider only nonredundant manipulators in this chapter, while the inverse kinematic problem of redundant manipulators will be included in Chapter 6.

The inverse kinematic problem is equivalent to searching for the solution to a set of nonlinear equations. Two concepts can be recognized. The first one is to determine the vectorial function $f^{-1}(x_e)$ in a symbolic form for each manipulator separately, assuming that the vectorial function $f(q)$ - has already been obtained in a symbolic form, and the second is to seek for the solution to nonlinear equations by some of the methods known from numeric analysis. Both these concepts have some advantages and some shortcommings. They will be discussed in detail in Sections 3.2 and 3.3. By using concrete examples we will see when it is possible and when it is impossible to obtain an explicit, analytical solution for the joint coordinates as functions of the end-effector position and orientation. We will also outline several numeric procedures for solving the inverse kinematic problem and discuss the problems encountered in the vicinity of manipulator singularities.

3.2. Analytical Solutions

Obtaining analytical solutions to inverse kinematic problems in different manipulators is rather a complex problem and it cannot be solved in a general way, in an arbitrary manipulator configuration. This is due to the fact that the vectorial function $f: q \rightarrow x_e$ is a relatively complex and nonlinear function in n variables. The analytical solution assumes that the joint coordinates are obtained as explicit functions in manipulator hand position and orientation. In the case when these functions can be obtained, then all the solutions $\{q^j, j \in J\}$ which satisfy the equation $f(q^j) = x_e$, are available. However, this is not the case with numeric algorithms for solving the inverse kinematic problem. Besides, the analytical solution is computationally more efficient then the corresponding numeric approach.

On the other hand, unlike the numeric solutions, the analytical solutions have to be derived by hand using a lot of intuition, for each manipulator separately [1, 19 - 22]. However, such manual solving of the inverse kinematic problem is feasible only for relatively simple kinematic configurations. One method of obtaining the analytical solution, which has been proposed in [1], will be illustrated later in this section.

Analytical solutions to the inverse kinematic problem for 6 degree-of--freedom open chain mechanisms where the axes of the three adjacent joints intersect at one point, have been considered in [23 - 24]. This restriction means that the problem can be separated into two independent subproblems: the calculation of the first three joint coordinates given the position of gripper base point, and the calculation of the remaining three coordinates given the orientation of the axes of the spherical joint. Both subproblems were reduced to finding the roots of a fourth degree polynomial in one unknown. In more complicated manipulator structures the degree of the corresponding polynomial rapidly increases (if it is possible to reduce the problem to a single polynomial at all [23]).

Let us now demonstrate the method proposed in [1] by several examples which will help us to ascertain whether or not it is possible to obtain the analytical solution.

Let us consider a simple 3 degree-of-freedom arthropoid manipulator as shown in Fig. 3.1. Denavit-Hartenberg kinematic parameters of the manipulator are given in Table T. 3.1.

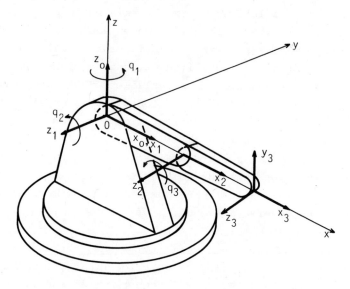

Fig. 3.1. 3 degree-of-freedom arthropoid manipulator

Link	Joint coordinate	α_i	a_i	d_i	$\cos\alpha_i$	$\sin\alpha_i$
1	q_1	90°	0	0	0	1
2	q_2	0°	a_2	0	1	0
3	q_3	0°	a_3	0	1	0

Table T.3.1. Denavit-Hartenberg parameters of the arthropoid manipulator

The adjacent homogeneous transformations are, according to Equation (1.3.25), given by

$$
{}^{0}T_{1} = \begin{bmatrix} C_1 & 0 & S_1 & 0 \\ S_1 & 0 & -C_1 & 0 \\ 0 & 1 & 0 & 0 \\ 0 & 0 & 0 & 1 \end{bmatrix} \tag{3.2.1}
$$

$$
{}^{1}T_{2} = \begin{bmatrix} C_2 & -S_2 & 0 & a_2 C_2 \\ S_2 & C_2 & 0 & a_2 S_2 \\ 0 & 0 & 1 & 0 \\ 0 & 0 & 0 & 1 \end{bmatrix} \tag{3.2.2}
$$

$$
{}^{2}T_{3} = \begin{bmatrix} C_3 & -S_3 & 0 & a_3 C_3 \\ S_3 & C_3 & 0 & a_3 S_3 \\ 0 & 0 & 1 & 0 \\ 0 & 0 & 0 & 1 \end{bmatrix} \tag{3.2.3}
$$

We will assume that the manipulator end-effector position and orientation have been given i.e. that the numeric values of the elements of the transformation matrix ${}^{0}T_{3} = {}^{0}T_{1}\,{}^{1}T_{2}\,{}^{2}T_{3}$ are known

$$
{}^{0}T_{3} = \begin{bmatrix} a_{11} & a_{12} & a_{13} & p_x \\ a_{21} & a_{22} & a_{23} & p_y \\ a_{31} & a_{32} & a_{33} & p_z \\ 0 & 0 & 0 & 1 \end{bmatrix} \tag{3.2.4}
$$

(Since we are dealing with a 3 degree-of-freedom manipulator in this example, we will here assume that only the position p_x, p_y, p_z is known). The joint coordinates will be determined by the following equations

$$
({}^{0}T_{1})^{-1}\,{}^{0}T_{3} = {}^{1}T_{3}
$$
$$
({}^{1}T_{2})^{-1}\,({}^{0}T_{1})^{-1}\,{}^{0}T_{3} = {}^{2}T_{3} \tag{3.2.5}
$$

The elements of matrices on the left-hand sides of the above equations depend on the elements of ${}^{0}T_{n} = {}^{0}T_{3}$ matrix and the joint coordinates $q_1,\ldots,q_i$, while the matrices on the right-hand sides depend on the remaining joint coordinates $q_{i+1},\ldots,q_n$. If a feasible solution exists, it is possible to obtain the analytical solution for the joint coordinates in a usefull form, by equating the elements of matrices on the left and the right-hand sides of the equations.

In the case of the given manipulator we have

$$(^{0}T_{1})^{-1} \, ^{0}T_{3} = \, ^{1}T_{3}$$

$$\begin{bmatrix} C_{1} & S_{1} & 0 & 0 \\ 0 & 0 & 1 & 0 \\ S_{1} & -C_{1} & 0 & 0 \\ 0 & 0 & 0 & 1 \end{bmatrix} \begin{bmatrix} a_{11} & a_{12} & a_{13} & P_{x} \\ a_{21} & a_{22} & a_{23} & P_{y} \\ a_{31} & a_{32} & a_{33} & P_{z} \\ 0 & 0 & 0 & 0 \end{bmatrix} =$$

$$= \begin{bmatrix} C_{2,3} & -S_{2,3} & 0 & a_{3}C_{2,3}+a_{2}C_{2} \\ S_{2,3} & C_{2,3} & 0 & a_{3}S_{2,3}+a_{2}S_{2} \\ 0 & 0 & 1 & 0 \\ 0 & 0 & 0 & 1 \end{bmatrix} \qquad (3.2.6)$$

The left-hand side of the above equation can also be presented in the form

$$(^{0}T_{1})^{-1} \, ^{0}T_{3} = \begin{bmatrix} f_{11}(a^{(1)}) & f_{11}(a^{(2)}) & f_{11}(a^{(3)}) & f_{11}(p) \\ f_{12}(a^{(1)}) & f_{12}(a^{(2)}) & f_{12}(a^{(3)}) & f_{12}(p) \\ f_{13}(a^{(1)}) & f_{13}(a^{(2)}) & f_{13}(a^{(3)}) & f_{13}(p) \end{bmatrix} \quad (3.2.7)$$

where $a^{(i)}$, $i=1,2,3$ and p denote the columns of matrix $^{0}T_{3}$, and

$$f_{11} = C_{1}x + S_{1}y$$

$$f_{12} = z \qquad\qquad (3.2.8)$$

$$f_{13} = S_{1}x - C_{1}y$$

Now, we are looking for the constant elements on the right-hand side of Equation (3.2.6) to correspond to a function of the unknown angle on the left-hand side. Here, by equating the elements 3, 4 on the left and right-hand sides, we obtain

$$S_{1}P_{x} - C_{1}P_{y} = 0 \qquad\qquad (3.2.9)$$

yielding

$$q_1 = \text{arctg} \frac{P_y}{P_x} + k\Pi \tag{3.2.10}$$

The next joint coordinate q_2 should be obtained by premultiplying by the next transformation i.e. from $(^1T_2)^{-1}(^0T_1)^{-1}\ ^0T_3 = {}^2T_3$. However, in this case the equation would not yield the solution for q_2, since the axes of the next two joints are parallel. Instead, we will equate the elements 1, 4 and 2, 4 on the left and right-hand sides of Equation (3.2.6)

$$C_1P_x + S_1P_y = a_3C_{2,3} + a_2C_2 \tag{3.2.11}$$

$$P_z = a_3S_{2,3} + a_2S_2$$

Since the angle q_1 has already been determined the left-hand sides of the above equations are known. By squaring both equations and then adding we obtain

$$C_3 = \frac{(C_1P_x+S_1P_y)^2 + P_z^2 - a_3^2 - a_2^2}{2a_2a_3} \tag{3.2.12}$$

This, in fact, corresponds to the cosine theorem. The S_3 variable is obtained from

$$S_3 = \pm\sqrt{1-C_3^2} \tag{3.2.13}$$

The signs + and - correspond to the elbow's up and down configuration. The angle q_3 is given by

$$q_3 = \text{arctg} \frac{S_3}{C_3} \tag{3.2.14}$$

Now, angle q_2 is obtained again from Equation (3.2.11) as

$$S_2 = \frac{(C_3a_3+a_2)P_z - S_3a_3(C_1P_x+S_1P_y)}{(C_3a_3+a_2)^2 + S_3^2a_3^2}$$

$$\tag{3.2.15}$$

$$C_2 = \frac{(C_3a_3+a_2)(C_1P_x+S_1P_y) + S_3a_3P_z}{(C_3a_3+a_2)^2 + S_3^2a_3^2}$$

$$q_2 = \text{arctg} \frac{(C_3a_3+a_2)P_z - S_3a_3(C_1P_x+S_1P_y)}{(C_3a_3+a_2)(C_1P_x+S_1P_y) + S_3a_3P_z} \tag{3.2.16}$$

This example shows that some rules may be established but that the analytical solving of the inverse kinematic problem requires a lot of intuition and special solutions for each manipulator. The method of sequential premultiplication by inverse transformations, facilitates the process of obtaining the solution in cases in which a solution exists, but an inflexible, straightforward application of the method would not always yield a solution. Some specific solutions must also be found out.

However, the explicit analytical solution for the joint coordinates as functions of the end-effector position and orientation does not always exist and it is only feasible in the kinematically simple manipulators, where it is possible to evaluate the joint coordinates one by one, independently. Let us illustrate this with the example of a simple 3 degree-of-freedom manipulator when it is not possible to obtain an explicit analytical solution for the joint coordinates.

Let us then consider an anthropomorphic manipulator with 3 revolute joints as shown in Fig. 3.2. Denavit-Hartenberg kinematic parameters of the mechanism are given in Table T.3.2.

The homogeneous transformations between the adjacent links for this manipulator are given by

$$
{}^{0}T_1 = \begin{bmatrix} C_1 & 0 & S_1 & a_1 C_1 \\ S_1 & 0 & -C_1 & a_1 S_1 \\ 0 & 1 & 0 & 0 \\ 0 & 0 & 0 & 1 \end{bmatrix}
\tag{3.2.17}
$$

$$
{}^{1}T_2 = \begin{bmatrix} C_2 & 0 & -S_2 & a_2 C_2 \\ S_2 & 0 & C_2 & a_2 S_2 \\ 0 & -1 & 0 & 0 \\ 0 & 0 & 0 & 1 \end{bmatrix}
\tag{3.2.18}
$$

$$
{}^{2}T_3 = \begin{bmatrix} C_3 & -S_3 & 0 & a_3 C_3 \\ S_3 & C_3 & 0 & a_3 S_3 \\ 0 & 0 & 1 & 0 \\ 0 & 0 & 0 & 1 \end{bmatrix}
\tag{3.2.19}
$$

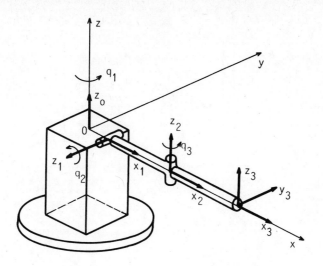

Fig. 3.2. Three degree-of-freedom anthropomorphic manipulator

Link	Joint coordinate	α_i	a_i	d_i	$\cos\alpha_i$	$\sin\alpha_i$
1	q_1	90°	a_1	0	0	1
2	q_2	-90°	a_2	0	0	-1
3	q_3	0	a_3	0	1	0

Table T.3.2. Denavit-Hartenberg parameters of the anthropomorphic manipulator

We will assume that the manipulator hand position, specified by p_x, p_y and p_z elements in matrix $^0T_3$, is known

$$^0T_3 = \begin{bmatrix} a_{11} & a_{12} & a_{13} & p_x \\ a_{21} & a_{22} & a_{23} & p_y \\ a_{31} & a_{32} & a_{33} & p_z \\ 0 & 0 & 0 & 1 \end{bmatrix}$$

Given the values p_x, p_y, p_z the joint coordinates q_1, q_2 and q_3 are to be determined. By applying the equation

$$(^0T_1)^{-1} \; ^0T_3 = \; ^1T_3$$

we obtain

$$
\begin{bmatrix}
C_1 & S_1 & 0 & -a_1 \\
S_1 & 0 & 1 & 0 \\
0 & -C_1 & 0 & 0 \\
0 & 0 & 0 & 1
\end{bmatrix}
\begin{bmatrix}
a_{11} & a_{12} & a_{13} & P_x \\
a_{21} & a_{22} & a_{23} & P_y \\
a_{31} & a_{32} & a_{33} & P_z \\
0 & 0 & 0 & 1
\end{bmatrix}
=
$$

$$
=
\begin{bmatrix}
C_3C_2 & -S_3C_2 & -S_2 & (a_3C_3+a_2)C_2 \\
C_3S_2 & -S_3S_2 & C_2 & (a_3C_3+a_2)S_2 \\
-S_3 & -C_3 & 0 & -a_3S_3 \\
0 & 0 & 0 & 1
\end{bmatrix}
\tag{3.2.20}
$$

By equating the right columns of the matrices on the left and right-
-hand sides of the above equation, we obtain

$$
C_1P_x + S_1P_y - a_1 = (a_3C_3+a_2)C_2
\tag{3.2.21}
$$

$$
P_z = (a_3C_3+a_2)S_2
\tag{3.2.22}
$$

$$
S_1P_x - C_1P_y = -a_3S_3
\tag{3.2.23}
$$

As opposed to the previous example, Equation (3.2.23) does not yield
the explicit solution for q_1, since there is not a constant, but a function of coordinate q_3 on the right-hand side of the equation. The system of Equations (3.2.21) - (3.2.23) has to be solved simultaneously in
this case. Equation (3.2.22) gives one of the relations between angles
q_2 and q_3, while another relation can be obtained by eliminating q_1
from Equations (3.2.21) and (3.2.23) (by squaring both equations and
adding)). By eliminating variables C_2 and S_2 from the equation obtained
and Equation (3.2.22), we obtain a fourth degree polynomial in the C_3
variable. This fully accords with the general results obtained in [23].
Thus the determination of the analytical solution to the inverse kinematic problem has been reduced to finding the roots of a fourth degree
polynomial in one unknown (e.g. by applying the Farrari method). This,
however, might be unsuitable from the computational standpoint (one
solution from the set of 4 solutions has to be selected). In more complicated kinematical manipulator configurations, the degree of the corresponding polynomial would rapidly increase, even if it were possible
to reduce the problem to that of finding the roots of a single polynomial.

One of the methods of obtaining the analytical solution for a given manipulator, often proposed in literature, is to simplify its kinematic structure by neglecting some of distance vectors (e.g. the end-effector was assumed to be a spherical joint although this is not often the case with real industrial manipulators). However, the analytical solution for the simplified manipulator structure would not yield good results in the motion synthesis of the real manipulator, since the approximation introduces a position and orientation error which, as a rule, cannot be overlooked.

A combination of the analytical and numerical methods for solving the inverse kinematic problem for an arbitrary 6 degree-of-freedom manipulator has been proposed in [107]. The solution is obtained as follows: first, some initial values for q_4, q_5 and q_6 are assumed and the first three joint coordinates are evaluated by solving the fourth order polynomial by applying the Farrari method. Then, coordinates q_4, q_5 and q_6 are calculated by taking the obtained values for q_1, q_2 and q_3 so as to improve the initial guess of q_4, q_5 and q_6. (Here the analytical solution is explicitly obtained since the gripper is assumed to be a spherical joint). The procedure is recursively repeated until the position and orientation errors are reduced under the given value. The procedure has been reported to have better convergence than in the Newton-Raphson algorithm [107].

3.3. Numerical Solutions

As we have seen, solving the inverse kinematic problem is equivalent to seeking for the solution to the set of nonlinear equations

$$q = f^{-1}(x_e) \tag{3.3.1}$$

$q \in Q \subset R^n$, $x_e \in X_e \subset R^n$. Therefore, the inverse kinematic problem can be obtained by applying some of the standard methods known from numeric analysis. These procedures are general and do not depend on mechanism configuration. However, as opposed to analytical methods numerical algorithms yield only one solution for the joint coordinates, given a vector of external coordinates x_e, which is closest to an initial guess q^o

$$q = f^{-1}(x_e, q^o) \tag{3.3.2}$$

In this section, we will first outline some of the numeric methods
which have been applied to the inverse kinematic problem solving. Later
we will point out the problem which arises in connection with the Jacobian matrix inversion. The following methods will be considered: the
Newton method [3-4, 25-26] which has been commonly used, and the Chebishev approach method proposed in [27]. Beside these methods, several
other procedures have been used in inverse kinematic problem solving:
gradient methods [28-29], coordinate descent methods [30-31], a method
based on dynamic programming [27], conjugate gradient methods, and so
on.

Beside purely analytical and purely numeric methods for obtaining the
inverse problem solution, a combination of these methods may be employed. In this case the Newton method is combined with the evaluation of
the symbolic Jacobian matrix, or, if possible, the Jacobian is inverted
analytically [21, 28, 32]. This problem will be discussed later in this
section.

The Newton method

This method has most frequently been used in solving the inverse kinematic problem for nonredundant (and redundant) manipulators. We will
outline it briefly here. Let us consider the function

$$F(q) = f(q) - x_e \tag{3.3.3}$$

where $q \in R^n$, $x_e \in R^n$ and $f: R^n \to R^n$ is a continuous, differentiable function
which maps joint coordinates into the external coordinates.

We should determine the zero of function F which is close to some approximate solution q^k. By expanding function $F(q)$ in Taylor's series
and keeping the first two terms only, we obtain

$$F(q) + J(q^k)(q-q^k) = 0 \tag{3.3.4}$$

where $J(q^k) = \partial f(q^k)/\partial q \in R^{n \times n}$ is the Jacobian matrix of partial derivatives of function f. Equation (3.3.4) is equivalent to

$$q = q^k - J^{-1}(q^k)(f(q^k)-x_e) = q^k + J^{-1}(q^k)\Delta x_e^k \tag{3.3.5}$$

where $J^{-1}(q^k)$ is inverse Jacobian matrix, $\Delta x_e^k = x_e-f(q^k)$ is the increment of external coordinates with respect to the external coordinates which correspond to the approximate solution q^k. If q is replaced by q^{k+1} in (3.3.5) an iterative procedure is obtained. The procedure should be ended when the condition $||\Delta x_e^k|| < \varepsilon$ is satisfied, where ε is a small positive constant. The Newton method has a quadratic convergency. It is clear that the Newton method yields a single solution for q, the solution which is closest to q^k.

However, manipulator motion is frequently set as a continuous trajectory in the external coordinates space, or by a set of points x_e^k, k=1,... ...,N, so that the increments $\Delta x_e^k = x_e^{k+1} - x_e^k$ are small. In that case the corresponding trajectory in the space of the joint coordinates is obtained by a single iteration of the Newton method

$$q^{k+1} = q^k + J^{-1}(q^k) \Delta x_e^k \qquad (3.3.6)$$

The solution q^{k+1} approximately satisfies $f(q^{k+1}) \simeq x_e^{k+1}$ and it is not necessary to check whether the error is less than ε, since it is assumed that the points x_e^k are sufficiently close to each other. Such a modification of the Newton-Raphson algorithm is very common in literature [1, 3, 7, 25].

However, this method of evaluating the trajectory in the space of the joint coordinates, given a trajectory of manipulator hand, inevitably leads to a certain accumulation the linearization error. In order to avoid this accumulation, a compromise between the original Newton--Raphson algorithm and Equation (3.3.6) was proposed in [33]. Namely, the joint coordinates are still evaluated according to (3.3.6) in a single iteration, but instead of evaluating Δx_e^k from $\Delta x_e^k = x_e^{k+1} - x_e^k$, the increment of external coordinates is corrected according to

$$\Delta x_e^k = x_e^{k+1} - f(q^k) \qquad (3.3.7)$$

In this case the evaluation of the trajectory in the space of the joint coordinates, given a trajectory of external coordinates, makes the evaluation of the Jacobian matrix and the real position vector $f(q^k)$ necessary at each sampling interval.

The equivalent to this is the relationship between the rates of external and joint coordinates, given a continuous trajectory $x_e(t)$

$$\dot{x}_e(t) = J(q(t))\dot{q}(t) \qquad (3.3.8)$$

Given the velocity vector $\dot{x}_e(t)$, one should determine the corresponding joint rates from

$$\dot{q}(t) = J^{-1}(q)\dot{x}_e(t) \tag{3.3.9}$$

This formulation of the inverse kinematic problem is obviously the equivalent of the Newton algorithm (3.3.5) as applied to an infinitesimal time interval.

A modification of Newton method

In the case when a manipulation task is specified by a set of points in the space of the external coordinates, which cannot be considered as close, the solution of the inverse kinematic problem is obtained by several iterations of the Newton algorithm (if the manipulator is not kinematically simple so that the analytical solution for joint angles is not available). The motion between these points is usually formed as joint-interpolated motion. It is desirable to speed up this process as much as possible.

One method of obtaining the inverse problem solution, i.e. finding the zeros of a nonlinear vector function was proposed in [26]. We will here outline the basic idea of this algorithm.

We will assume that an initial vector x_e^O and the corresponding vector q^O, which satisfies $f(q^O) = x_e^O$, $x_e^O \in x_e \in R^n$, $q \in Q \in R^n$ are given, together with the final vector of external coordinates x_e^F. The vector of joint coordinates which satisfies

$$f(q^F) = x_e^F \tag{3.3.10}$$

is to be determined. The modification of the Newton method involves the optimization of a parameter $s \in (0, 1]$, defined by

$$\Delta x_e^k = s(x_e^F - x_e^k)$$

at each iteration k. The optimal value $s = s^*$ in inertion k is obtained by minimizing the performance index

$$C = [x_e^F - x_e(s)]^T [x_e^F - x_e(s)] \tag{3.3.11}$$

The value of the external coordinates vector $x_e(s)$ in iteration k is given by

$$x_e(s) = f(q^k + s\Delta q) \qquad (3.3.12)$$

where

$$\Delta q = J^{-1}(q^k)(x_e^F - x_e^k), \qquad x_e^k = f(q^k) \qquad (3.3.13)$$

According to the modified Newton-Raphson algorithm (3.3.12) - (3.3.13) and evaluating the optimal value s^* with respect to (3.3.11), we obtain $x_e^{k+1} = x_e(s^*)$, Fig. (3.3).

An approximate evaluation of the minimum of (3.3.11), according to the orthogonality condition, was proposed in [26]

$$[x_e(s^*) - x_e^F]^T \frac{dx_e(s)}{ds}\bigg|_{s=s^*} = 0 \qquad (3.3.14)$$

This equation is a necessary condition for the minimum. According to this procedure s is evaluated iteratively from

$$s_{i+1} = s_i + r \frac{[x_e^F - x_e(s_i)]^T [x_e(s_i) - x_e(s_{i-1})]}{|x_e^F - x_e(s_i)||x_e(s_i) - x_e(s_{i-1})|} , \quad i=0,1,\ldots \qquad (3.3.15)$$

where s_o and r have to be chosen. The iterative procedure should be ended when the value of the fraction in (3.3.15) falls under a specified value (e.g. 0.1 or 0.2). However, examples showed that minimization according to (3.3.15) may produce anomalies, when s_{k+1} yields a value of C greater then the value obtained for s_k. This is a consequence of the fact that several values of s exist, which satisfy (3.3.14). Practical examples, however indicate that it is sufficient to limit only the maximal number of iterations, without searching for special methods to overcome these problems.

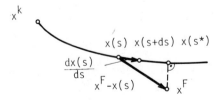

Fig. 3.3. An illustration of the modified Newton algorithm

The results of experiments [26] showed that the inverse problem solving is significantly accelerated. In a six degree-of-freedom manipulator the average number of iterations was 4-6 (with a maximum 8 for $\Delta q=120°$). Each iteration requires approximately 4 more steps to determine s^*, but this is less time consuming as it requires only the evaluation of $f(q)$, without $J^{-1}(q)$.

The Chebishev approach method

This method was proposed in [27]. The problem is to determine such a solution q^* to the system of equations

$$J(q)\dot{q} - \dot{x}_e = \varepsilon(\dot{q}) \tag{3.3.16}$$

where $\varepsilon = [\varepsilon_1 \cdots \varepsilon_n]^T$, which satisfies

$$\max_i |\varepsilon_i(\dot{q}^*)| = \min_q \max_i |\varepsilon_i(q)| = L^* \tag{3.3.17}$$

Point $\dot{q}^*$ is geometrically closest to the hyperplane (3.3.8), so that in Chebishev sense vector $\dot{x}_e^* = J\dot{q}^*$ is closest to the given vector $\dot{x}_e$.

The Chebishev approach to the solution is reduced to the problem of linear programming, so that it is appropriate to take the constraints on joint velocities into account

$$|\dot{q}_i| < \alpha_i, \qquad i=1,\ldots,n \tag{3.3.18}$$

together with all other constraints on the mechanism motion. The linear programming technique applied to finding Chebishev approach to the solution to (3.3.8), with constraints (3.3.18), has the following form: it is necessary to find the negative values of variables $x_1,\ldots,x_{n+1}$ which minimize linear function $z = x_{n+1}$, with constraints

$$\sum_{j=1}^{n} J_{ij}x_j - x_{n+1} < b_i$$

$$\sum_{j=1}^{n} J_{ij}x_j + x_{n+1} > b_i, \qquad x_i < 2\alpha_i, \qquad i=1,\ldots,n \tag{3.3.19}$$

where $b_i = \sum_{j=1}^{n} J_{ij}\alpha_j + \dot{x}_e$, $x_i = \dot{q}_i + \alpha_i$, J_{ij} is the i, j element of matrix J. This problem is efficiently solved by the dual simplex method.

The inverse Jacobian

As we have observed the numeric method of solving the inverse kinematic problem requires the evaluation of the Jacobian matrix. This subject has been considered in detail in Chapters 1 and 2. The Newton method also requires the solution of the linear set of n equations in n unknowns

$$\Delta x_e = J(q) \Delta q \qquad (3.3.20)$$

$\Delta x_e \in R^n$, $\Delta q \in R^n$, $J(q) \in R^{n \times n}$, or equivalently,

$$\dot{x}_e = J(q) \dot{q} \qquad (3.3.21)$$

which is to be solved in the rate control algorithms. Two main questions may be set. The first is the method of solving the system, and the second is the existence of the solution.

Let us first consider the possibility of obtaining an analytical solution for joint rates given the velocity of external coordinates. If, given a manipulator configuration, the Jacobian is obtained in a symbolic form, it may be possible to invert it symbolically [34]. However, such inversion is usually very difficult due to the complexity of the Jacobian itself.

On the other hand, if it is possible to obtain joint coordinates as the explicit functions of the external coordinates vector x_e and the already evaluated joint coordinates, the joint velocities $\dot{q}$ can be obtained by directly differentiating these relations [1, 21]. This solution is required in the case of rate control algorithms. However, as we have seen in Section 3.2, this is feasible only for kinematically simple manipulators.

In [28, 32] the joint rates have also been obtained as the explicit functions of the linear and angular velocities of the end-effector of a 6 degree-of-freedom manipulator with a spherical joint in the gripper. The spherical joint made it possible for the rates $\dot{q}_1$, $\dot{q}_2$ and $\dot{q}_3$ to be evaluated as the functions of the linear velocity of the gripper base, apart from the remaining three joint rates $\dot{q}_4$, $\dot{q}_5$, $\dot{q}_6$, which are obtained as functions of the angular velocity of the gripper.

Most frequently, however, the solution to the linear system of equations (3.3.21) is obtained by applying some of the methods known from numeric analysis (Jacobi method, Gauss-Seidel algorithm, and so on) [35]. Iterative methods are however, inconvenient from the standpoint of real-time application. Direct methods for obtaining the solution to a linear system of equations (such as Gauss elimination and its various modifications) are, therefore, most often used.

However, the process of solving the inverse kinematic problem is complicated by singularities which occur whenever the manipulator becomes degenerate. Then the Jacobian matrix $J \epsilon R^{n \times n}$ becomes singular and rankJ< <n, det J=0. Here we come to the question of the existence of the solution, which is defined by the Cronecker-Capelli theorem [35]:

1. if matrix J(q) is a full rank matrix, i.e. rank J=n, system (3.3.21) has a unique solution;

2. if rank J<n and rank J = rank$[J \mid \dot{x}_e]$ an unbounded number of solutions exist;

3. if rank J $\neq$ rank$[J \mid \dot{x}_e]$ no solution to system (3.2.21) exists.

In motion generation, singularities are usually avoided by modifying the trajectory in the space of the external coordinates in the vicinity of singularities [19]; the other method is to form a joint-interpolated motion in the vicinity of singularities.

Instead of avoiding the problem of singularities, there is a possibility of selecting one of the solutions from the unbounded set of solutions. In the case when some of the rows of J are mutually dependent it is possible to obtain a solution to (3.3.20) by striking out the colinear rows; the obtained matrix $J_p \epsilon R^{n \times n}$ is nonsingular, rank J_p=m. The minimum norm solution to the underdetermined system is now given by

$$\Delta q = J_p^T (J_p J_p^T)^{-1} \Delta x_e, \qquad m<n, \text{ rank } J_p = m \qquad (3.3.22)$$

In the other case, when the columns of J are not independent, the solution is obtained by the right generalized inverse of the nonsingular submatrix of J, $J_p \epsilon R^{n \times m}$

$$\Delta \dot{q} = (J_p^T J_p)^{-1} J_p^T \Delta x_{e'}, \qquad n>m, \text{ rank } J_p = m \qquad (3.3.23)$$

where J_p is obtained by striking out the colinear columns.

However, the problem lies in determining the nonsingular submatrices
of J. Numerical methods are time consuming [36]. However, given a manipulator and it's symbolic Jacobian matrix, the singularities are known
in advance and this information can be utilized in determining which
columns or rows are linearly dependent (or come very close to linear
dependance).

If the linear system of equations (3.3.20) was solved by applying the
singular-value decomposition method [37], explicit determination of the
nonsingular submatrices of J would not be needed. However, the time required for evaluating the pseudoinverse by this method is about 15 times
longer than when applying (3.2.22) [36]. Besides, in the vicinity of
some singularities the pseudoinverse control may, also yield unacceptably high joint rates [103]. This method and it's application to the
inverse kinematic problem of redundant manipulators will be discussed
in Chapter 6 (Section 6.2).

Methods of overcomming the problems of singularities in motion synthesis will certainly represent a subject of further investigation.

3.4. Summary

We have considered the inverse kinematic problem in this chapter. Two
main methods for obtaining joint coordinates, given the position and
orientation of manipulator end-effector, have been discussed. The first
refers to analytic methods, where joint angles are expressed as the
explicit functions of the external coordinates and the already evaluated joint coordinates. Such a solution, when it can be derived, is
the least complex in computation, but it is only feasible for kinematically simple manipulators. On the other hand, simplification of the
kinematic structure of a given robot, in order to obtain an analytic
solution, would result in significant position and orientation errors.

On the other hand, numeric methods are computationally more complex and
consequently take more time. Besides, the singular points, where the manipulator becomes degenerate complicate the application of numerical
methods in motion generation.

Chapter 4
Kinematic Approach to Motion Generation

4.1. Introduction

Having solved the direct and the inverse kinematic problem, the primary problem arising in robot control is manipulator motion generation.
Methods of specifying manipulator motion, i.e. the description of a
concrete manipulation task, significantly affects manipulator flexibility
and its usefulness in various industrial applications.

The basic property, which the motion generation algorithms should pos-
sess is provision for the functionality of motion, i.e. to provide for
continuity and smoothness of motion through the required gripper positions and orientations. Beside this basic requirement, additional properties are also needed, to increase the adaptivity and flexibility of
robots. For example, motion generation algorithms should provide for
motion synthesis in environment with obstacles, the treatment of various types of sensors, such as cameras, tactile sensors, force sensors,
proximity sensors, working on moving conveyors, and so on.

In this book we will consider the methods used to date, in order to re-
solve the basic problems in motion generation, i.e. we will discuss
methods of specifying trajectories between a set of given points in the
space of the external coordinates. We will not here consider the robot
programming languages, which have been developed to simplify and facilitate the process of teaching robots how to perform specific manipulation tasks [37 - 46]. These languages should, beside achieving desired positions and orientations, also provide a broader set of faci-
lities, such as commands referring to the gripping function, synchro-
nization with other processes, control of forces in contact with the
work object, etc. We will, instead, here present the mathematical basis
for motion generation algorithms.

In Section 4.2. the manipulation task will be formulated. The problem
of trajectory planning in the sense of specifying the set of points in
the space of the external coordinates where the end-effector should
pass, will be briefly discussed in Section 4.3. Motion generation be-

tween the given set of points will be considered in Section 4.4. This synthesis can be performed either in the space of the joint coordinates, or in the space of the external coordinates. The engineering procedures, which have been most frequently used in robot motion synthesis, will be presented, as well as some of the methods proposed in literature, which take into account some kinematic performance index. These discussions do not, in any case, consider the dynamic model of the robot. Robot motion synthesis which is based on the dynamic model of the system will be examined in Chapter 5. Procedural motion synthesis, employed when a trajectory is geometrically defined (such as circles, ellipses, etc.), will be discussed in Section 4.5.

4.2. Manipulation Task

In this section we will consider the definition of a manipulation task from the kinematic point of view only, i.e. the manipulation task specification will be considered as the problem of specifying the desired positions and orientations of the gripper. The problem of motion execution in the presence of various disturbances will not be considered.

Generally, a manipulation task can be specified by an arbitrary, continuous trajectory in the space of the external coordinates versus time $x_e(t) \in X_e$, $\forall t \in [0, T]$, where T is movement execution time. The manipulator end-effector should move along this trajectory during task execution. The method of defining manipulation tasks, where external coordinates vector (positions and orientations) are imposed is termed positional motion synthesis. On the other hand, motion synthesis where velocities $\dot{x}_e(t)$ or accelerations $\ddot{x}_e(t)$ are specified, together with the initial position and velocity is termed "rate manipulator control". Rate control is employed with "master-slave" manipulators, during the process of teaching the robot which points to pass through.

Manipulation task specification usually involves two stages:

1. choosing the set of points $\{x_e^0, x_e^1, \ldots, x_e^N\}$ in the space of the external coordinates, which the manipulator end-effector should pass trough in time instants $\{t_0, t_1, \ldots, t_N = T\}$;

2. selecting motion trajectories between points x_e^k, $k=0, \ldots, N$.

The first stage will be termed as "motion planning" in the text to follow, while the second phase will be termed "motion synthesis".

Instead of specifying a set of locations in space, it is sometimes convenient to generate procedural motions as continuous trajectories $x_e(t)$.
This is suitable for geometrically defined tasks, such as circles, elliptic paths, etc. Procedural motions will be discussed in Subsection
4.4.2.

Another way of defining a manipulation task is tabular motion specification, used when direct teaching methods are employed in activities
such as defining spray painting tasks.

4.3. Trajectory Planning

At the stage of trajectory planning we should first select the type of
external coordinates vector, which will describe the manipulation task.
This decision depends on the manipulation task itself, as well as the
kinematical configuration of the mechanism.

The selection of points x_e^k, $k=0,\ldots,N$ is mostly determined by technological requirements, the distribution of handled objects, tools, obstacles, etc. The time instants t_k, $k=1,\ldots,N$ are usually selected em-
pirically, compromising between technological requirements and estimated maximal velocity and acceleration of the joints. Instead of defining time instances t_k, desired end-effector velocities may also be
specified, while the time instances are evaluated so that none of the
joints reach their maximal speed. On the other hand, t_k allocation may
also become a subject of optimization with respect to some kinematic or
dynamic performance index [47 - 49].

Specification of positions and orientations at trajectory end points
can be performed in several ways. One of them is to store joint coordinates in the process of teaching by doing, where an operator moves
the manipulator end-effector along a desired path (e.g. for painting
robots). Another way is to store the points where the manipulator is
brought by means of a teaching box, or the points are defined by a robot language commands.

The end points of trajectory segments are usually specified with respect to a reference coordinate system attached to the manipulator base.
On the other hand, it is possible to describe the manipulation task with

respect to several coordinate frames attached to various tools and objects in manipulator work space [1, 50 - 52]. This more general method of trajectory specification is termed structured robot programming and teaching. It is convenient from the standpoint of an operator defining the task, since it is easier to specify hand position and orientation with respect to a tool instead of a distant coordinate frame attached to manipulator base. It requires greater computational complexity, since before solving the inverse kinematic problem all the end points have to be transformed into the same coordinate frame. If one is dealing with on-line motion generation, considerable computer capabilities are then required. However, the advantages of the structured task description will certainly become apparent thanks to fast computer technology development.

Manipulator motion planning in the presence of obstacles is also an important aspect of robot motion generation, and this is now being intensively developed [53 - 56]. We are not going to consider this problem as the problem of searching for a free end-effector trajectory between obstacles in this book. Collison-free motion synthesis for redundant manipulators, in the case when end-effector path is given, will be discussed in Chapter 6.

4.4 Motion Between Positions

The second stage of trajectory generation involves interpolation between the specified set of points x_e^k, $k=0,\ldots,N$. This problem can be considered as:

1. interpolation in the space of the external coordinates between points x_e^{k-1} and x_e^k, $k=1,\ldots,N$;

2. interpolation in the space of the joint coordinates (joint-interpolated motion), where the trajectory $q(t)$ between values $q^{k-1} = f^{-1}(x_e^{k-1})$ and $q^k = f^{-1}(x_e^k)$ is specified.

Various approaches to motion synthesis exist both for joint-interpolated and Cartesian motion. In methodological terms, it is possible to search for the solution satisfying certain functional requirements and constraints, with no attention paid to optimality in any sense. On the other hand, it is possible to search for an optimal solution with respect to a performance index.

Both suboptimal and optimal motion synthesis should take constraints
into account. The kinematical methods of motion synthesis, which we are
dealing with in this chapter, involve the following types of constraints:

1. constraints on joint coordinates resulting from mechanical structure
   of the mechanism;

2. constraints on maximal velocities and accelerations of joint coordinates;

3. constraints on joint coordinates imposed by external obstacles.

If the manipulator end-effector path is considered as an arbitrary,
continuous trajectory in the space of the external coordinates, it is
obvious that it is possible to provide for continuity of velocity and
acceleration, too. This is rather important in industrial applications,
where smooth, jerkless motion is preferable. However, if the task specification is considered as a two stage procedure, where points x_e^k,
k=0,...,N are specified, and then the motion between these positions is
generated, only the continuity of position (the external coordinates)
is explicitly provided. By imposing additional constraints on external
coordinates velocities and acceleration in the user-specified points
it is possible to ensure the continuous variation of both joint velocity and acceleration. However, instead of interpolating a high order
polynomial between these points, it is easier to form straight line
segments between these end points and to set velocity to zero at these
points. This method has been most frequently employed in robot motion
generation.

In the tasks where it is sufficient to pass smoothly, close enough to
the transition points x_e^k, approximately in time instants t_k, it is possible to synthesize trajectories [1, 52], which provide for the continuity of position, velocity and acceleration. Thus, the problem of motion
synthesis is not only regarded as a set of practically nonrelated subproblems each associated to one trajectory segment x_e^k, x_e^{k+1}, k=0,...,N-1,
or equivalently q^k, q^{k+1} if joint-interpolated motion is considered.

In the text to follow we will give a survey of these methods for motion
synthesis. We will first consider joint-interpolated motion and afterwards Cartesian motion. Once the time schedules $x_e(t)$ are specified,
one should determine the corresponding joint coordinates trajectory
$q(t)$ and $\dot{q}(t)$ by solving the inverse kinematic problem. The trajectories

obtained are also termed nominal trajectories in the kinematical sense.

4.4.1. Joint-interpolated motion

Once the end points of trajectory segments x_e^k, $k=0,\ldots,N$ and the time instances t_k are specified, the corresponding joint coordinates q^k, $k=0,\ldots,N$ are evaluated by solving the inverse kinematic problem $q^k = = f^{-1}(x_e^k(t_k))$. Joint-interpolated motion assumes that time schedules $q(t)$, $t_{k-1} \leqslant t \leqslant t_k$, $k=1,\ldots,N$ will be specified directly. This motion synthesis does not provide for the straight line motion of the end-effector in the manipulator work space, since the linear interpolation is performed in the joint coordinates space. However, this method of motion synthesis is computationally far more efficient than Cartesian motion, since it does not require inverse problem solving along the trajectory.

We will first consider some engineering methods for the suboptimal forming of joint coordinates trajectories consisting of N segments. Each segment is formed independently of the others. Here we will turn to some methods of motion synthesis in the joint coordinates space, where two or more trajectory segments are treated simultaneously.

Motion between two points

For the sake of simplicity, we will denote points q^k and q^{k+1} by q^O and q^F, respectively, and assume that the motion execution time is T ($t \in [0, T]$). One of the most common methods of motion synthesis is linear interpolation between the initial and the final point, according to a given velocity profile

$$q(t) = q^O + \lambda(t)(q^F - q^O), \qquad 0 \leqslant t \leqslant T \qquad (4.4.1)$$

or

$$\dot{q}(t) = \dot{\lambda}(t)(q^F - q^O), \qquad 0 \leqslant t \leqslant T \qquad (4.4.2)$$

where $\lambda(t) \in [0, 1]$ is a scalar parameter specifying velocity distribution along the trajectory. One can choose from various time schedules for $\lambda(t)$, which result in various velocity profiles $\dot{q}(t)$, e.g. constant velocity, triangular, trapezoidal, parabolic, cosine profile, etc., (Fig. 4.1).

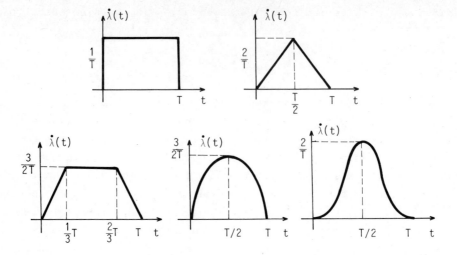

Fig. 4.1. Manipulator velocity profiles

Constant velocity profile $\dot{\lambda}(t) = \frac{1}{T}$ is not suitable for practical applications, owing to the existence of peaks in acceleration at initial and final time instants. The triangular profile is a special case of the symmetrical trapezoidal profile

$$\dot{\lambda}(t) = \begin{cases} \dfrac{4t}{T^2}, & 0 < t < \dfrac{T}{2} \\[2mm] \dfrac{4(T-t)}{T^2}, & \dfrac{T}{2} < t < T \end{cases} \qquad (4.4.3)$$

The trapezoidal velocity profile, having the acceleration and deceleration time $\frac{T}{3}$, is given by

$$\dot{\lambda}(t) = \begin{cases} \dfrac{9t}{2T^2}, & 0 < t < \dfrac{1}{3}T \\[2mm] \dfrac{3}{2T}, & \dfrac{1}{3}T < t < \dfrac{2}{3}T \\[2mm] \dfrac{9(T-t)}{2T^2} & \dfrac{2}{3}T < t < T \end{cases} \qquad (4.4.4)$$

This velocity profile is most commonly used in industrial robot motion synthesis.

The parabolic velocity profile, given by

$$\lambda(t) = \frac{6}{T^2}\left(\frac{1}{2} - \frac{1}{3}\frac{t}{T}\right)t^2$$

$$\dot{\lambda}(t) = \frac{6}{T^2}(1 - \frac{t}{T})t, \qquad 0 < t < T \qquad (4.4.5)$$

$$\ddot{\lambda}(t) = \frac{6}{T^2}(1 - 2\frac{t}{T})$$

has an advantage over the trapezoidal profile, since it has no abrupt changes of acceleration along the trajectory, except at the initial and final time instant.

The cosine velocity profile satisfies $\dot{\lambda}(0) = \dot{\lambda}(T) = \ddot{\lambda}(0) = \ddot{\lambda}(T) = 0$. It is given by

$$\lambda(t) = \frac{t}{T} - \frac{1}{2\Pi} \sin \frac{2\Pi}{T} t$$

$$\dot{\lambda}(t) = \frac{1}{T}(1 - \cos \frac{2\Pi}{T} t), \qquad 0 < t < T \qquad (4.4.6)$$

$$\ddot{\lambda}(t) = \frac{2\Pi}{T^2} \sin \frac{2\Pi}{T} t$$

The cosine profile is computationaly more complex than the trapezoidal and parabolic profile.

A polynomial function which also satisfies $\dot{\lambda}(0) = \dot{\lambda}(T) = \ddot{\lambda}(0) = \ddot{\lambda}(T) = 0$ and $\dddot{\lambda}(0) = \dddot{\lambda}(T) = \text{const.}$, is the fifth order polynomial

$$\lambda(t) = 10(\frac{t}{T})^3 - 15(\frac{t}{T})^4 + 6(\frac{t}{T})^5$$

$$\dot{\lambda}(t) = \frac{1}{T}[30(\frac{t}{T})^2 - 60(\frac{t}{T})^3 + 30(\frac{t}{T})^4]$$

$$\ddot{\lambda}(t) = \frac{1}{T^2}[60(\frac{t}{T}) - 180(\frac{t}{T})^2 + 120(\frac{t}{T})^3] \qquad (4.4.7)$$

$$\dddot{\lambda}(t) = \frac{1}{T^3}[60 - 360(\frac{t}{T}) + 360(\frac{t}{T})^2]$$

It is evident that as the number of terminal conditions increases, i.e. the smoother the trajectory required, so the computational complexity also increases.

The joint velocity profiles $\dot{q}(t)$ are proportional to the above profiles $\dot{\lambda}(t)$, and the motion is performed along the straight line segment $q^O - q^F$ in the space of the joint coordinates. This motion clearly deviates from the straight line motion in the manipulator work space. The magnitude of the deviation depends on the distance between points $x_e^O = f(q^O)$ and $x_e^F = f(q^F)$.

All these velocity profiles $\dot{\lambda}(t)$ can also be applied to motion synthesis in the space of the external coordinates, and therefore will be considered in more detail in Subsection 4.4.2.

Suboptimal motion synthesis in the space of the
joint coordinates

Let us now consider motion synthesis between a set of points $q^k = f^{-1}(x_e^k)$, $k=0,\ldots,N$, through which the manipulator should pass at time instants t_k, and which provides for continuity of velocity and acceleration in the transition points q^k. Several methods were proposed in literature [1, 3, 19, 47, 48, 52] refering to joint-interpolated motion. We will here outline some of these methods.

In order to avoid abrupt changes of velocity and acceleration in the transition points, R. Paul proposed a quadratic change of acceleration $q(t)$ in the vicinity of the transition points, which applies to the tasks where it is sufficient to pass close enough to the point [1, 47] (Fig. 4.2). According to the notation introduced in Fig. 4.2, the time

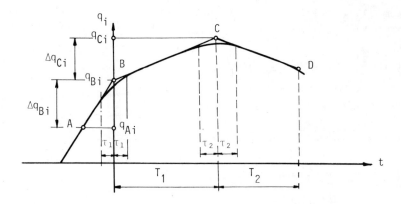

Fig. 4.2. Time history of joint coordinate q_i

history of $q(t)$, $\dot{q}(t)$, $\ddot{q}(t)$ in the time interval $-\tau_1 \leqslant t \leqslant \tau_1$, can be expressed as

$$q = [(\Delta q_C \frac{\tau_1}{T_1} - \Delta q_B)(2-h)h^2 + 2\Delta q_B]h + q_A$$

$$\dot{q} = [(\Delta q_C \frac{\tau_1}{T_1} - \Delta q_B)(1.5-h)2h^2 + \Delta q_B]\frac{1}{\tau_1}$$

$$(4.4.8)$$

$$\ddot{q} = (\Delta q_C \frac{\tau_1}{T_1} - \Delta q_B)(1-h)\frac{3h}{\tau_1^2}$$

$$h = \frac{t+\tau_1}{2\tau_1}, \qquad -\tau_1 < t < \tau_1$$

while for the intermediate part of the trajectory the velocity is constant

$$q(t) = \Delta q_C \frac{t}{T_1} + q_B, \qquad \tau_1 < t < T_1 - \tau_2 \qquad (4.4.9)$$

Here, six boundary conditions are imposed (for position, velocity and acceleration in both end points), which makes it necessary for sixth order polynomial to be interpolated. However, the fourth order polynomial is sufficient, owing to the symmetry.

The execution times T_k are defined by maximal stationary velocity and the increment Δq ($T = \max\limits_{i=1,\ldots,n} \Delta q_i/v_i$). The constants τ_k are estimated from

$$\tau = \frac{3}{4} \frac{\Delta v}{a}$$

where Δv is the increment of velocity between the two segments, and a is the maximal acceleration allowed.

Thus the continuity of position, velocity and acceleration is provided for; however, a position error in the transition points is simultaneously introduced. The time function $q(t)$, $-\tau_1 < t < T_1 - \tau_2$ depends solely on the variables related only to the three points A, B and C on the trajectory, and does not depend on the other trajectory segments. This feature is convenient in real time motion synthesis.

A further generalization of the above described smooth transition from one path segment to another was proposed in [52]. It provides not only for the continuity of position, velocity and acceleration, but also for continuity of jerk (the third derivative). The resultant polynomial is of the seventh order, being computationally rather complex.

A.F. Vereshchagin and V.L. Generozov made use of cubic spline functions in to order to ensure continuity of $q(t)$, and its first and second derivative on the entire time interval [0, T] between two end-effector stops [19]. The third order spline function, which determines the joint coordinate q_i between points q_i^k and q_i^{k+1}, $i=1,\ldots,n$, $k=0,\ldots,N-1$ is given by

$$q_i(t) = \omega_i^k \frac{(t_{k+1}-t)^3}{6h_k} + \omega_i^{k+1} \frac{(t-t_k)^3}{6h_k} + (q_i^k - \frac{\omega_i^k h_k^2}{6})\frac{t_{k+1}-t}{h_k} +$$

$$+ (q_i^{k+1} - \frac{\omega_i^{k+1} h_k^2}{6})\frac{t-t_k}{h_k}, \qquad t_k < t < t_{k+1} \qquad (4.4.11)$$

where $h_k = t_{k+1}-t_k$. This function satisfies boundary conditions

$$q_i(t_k) = q_i^k, \qquad k=0,\ldots,N \qquad i=1,\ldots,n$$

$$\dot{q}_i(t_o) = \dot{q}_i(T) = 0, \qquad i=1,\ldots,n \qquad (4.4.12)$$

Variables $\omega_i^k = \ddot{q}_i(t_k)$ are evaluated as the solution to the following linear system of equations

$$2\omega_i^o + \omega_i^1 = 6 \frac{q_i^1-q_i^o}{h_o^2}$$

$$h_{k-1}\omega_i^{k-1} + 2(h_k+h_{k-1})\omega_i^k + h_k\omega_i^{k+1} = 6[\frac{q_i^{k+1}-q_i^k}{h_k} - \frac{q_i^k-q_i^{k-1}}{h_{k-1}}],$$

$$\omega_i^{N-1} + 2\omega_i^N = -6 \frac{q_i^N-q_i^{N-1}}{h_{N-1}^2} \qquad k=1,\ldots,N-1 \qquad (4.4.13)$$

Once the ω_i^k variables are evaluated, it is necessary to check whether the constraints

$$\max_{t_o<t<T} |\dot{q}_i(t)| < v_i, \qquad \max_{t_o<t<T} |\ddot{q}_i(t)| < a_i, \qquad i=1,\ldots,n \qquad (4.4.14)$$

are satisfied. If these constraints are not satisfied, the authors propose that the parameters h_k and ω_i^k are recalculated according to

$$\tilde{h}_k = h_k\sqrt{\lambda}, \qquad k=0,\ldots,N-1, \qquad \tilde{\omega}_i^k = \omega_i^k/\lambda, \qquad i=1,\ldots,n, \ k=0,\ldots,N$$

$$\lambda = \max(1, \lambda_1^2, \lambda_2)$$

$$\lambda_1 = \max_i[v_i^{-1}(\max_i \max_{t\in[t_k,t_{k+1}]} |\dot{q}_i(t)|)] \qquad (4.4.15)$$

$$\lambda_2 = \max_i[a_i^{-1}(\max_k|\omega_i^k|)]$$

Thus, the joint-interpolated motion is obtained, which satisfies the condition (4.4.12) and (4.4.14).

Cubic spline functions have also been used in [57] in order to avoid overshooting or undershooting in the user-specified positions, while not reducing the execution speed by setting the velocity to zero if this is not really necessary. Here the acceleration profile is adopted to be trapezoidal (Fig. 4.3), with the rise time being constant and equal $\frac{1}{16}$ of the total execution time between two positions. According to this procedure, given the initial position q^o and velocity v^o, the final position q^1 and velocity v^1 and the execution time $16T$, the joint coordinate q is evaluated from the following set of equations, each for the six regions

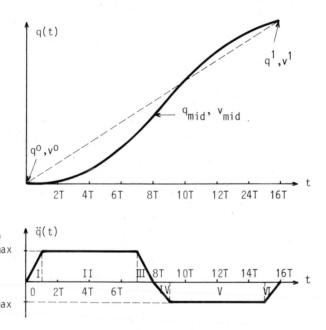

Fig. 4.3. Time history of a joint coordinate and its acceleration

$$q(t) = \frac{1}{6T} a^o_{max} t^3 + v^o t + q^o, \qquad 0 < t < T$$

$$q(t) = \frac{1}{2} a^o_{max} t^3 + (v^o - \frac{1}{2} T a^o_{max})t + q^o + \frac{1}{6} T^2 a^o_{max}, \qquad T < t < 7T$$

$$q(t) = -\frac{1}{6T} a^o_{max} t^3 + 4 a^o_{max} t^2 + (v^o - 25 T a^o_{max})t +$$
$$\quad + q^o + \frac{172}{3} T^2 a^o_{max}, \qquad 7T < t < 8T$$

$$q(t) = \frac{1}{6T} a^1_{max} t^3 - 4 a^1_{max} t^2 + (v_{mid} + 32 T a^1_{max})t +$$
$$\quad + q_{mid} - 8T v_{mid} - \frac{256}{3} T^2 a^1_{max}, \qquad 8T < t < 9T$$

$$q(t) = \frac{1}{2}a_{max}^1 t^2 + (v_{mid} - \frac{17}{2}Ta_{max}^1)t +$$

$$+ q_{mid} - 8Tv_{mid} + \frac{217}{6}T^2 a_{max}^1, \qquad 9T \leqslant t \leqslant 15T$$

$$q(t) = -\frac{1}{6T}a_{max}^1 t^3 + 8a_{max}^1 t^2 + (v_{mid} - 121Ta_{max}^1)t +$$

$$+ q_{mid} - 8Tv_{mid} + \frac{1796}{3}T^2 a_{max}^1 \qquad (4.4.16)$$

The quantities q_{mid}, v_{mid} denote a midpoint and its velocity. They are determined from the condition that the position and velocity at the final point should have the values q^1 and v^1

$$v_{mid} = v^1 - 7Ta_{max}^1$$

$$q_{mid} = q^1 - 8Tv^1 + 28T^2 a_{max}^1 \qquad (4.4.17)$$

Maximum accelerations for the two path segments are evaluated by ensuring that the first half of the trajectory terminates at q_{mid} with velocity v_{mid}

$$a_{max}^o = \frac{q^1 - q^o}{56T^2} - \frac{v^1 + 3v^o}{14T}$$

$$a_{max}^1 = \frac{q^o - q^1}{56T^2} + \frac{3v^1 + v^o}{14T} \qquad (4.4.18)$$

Substituting (4.4.18) into (4.4.17) one obtains

$$v_{mid} = \frac{q^1 - q^o}{8T} - \frac{1}{2}(v^1 + v^o)$$

$$q_{mid} = \frac{q^o + q^1}{2} + 2T(v^o - v^1) \qquad (4.4.19)$$

If there are several trajectory segments, the velocity v^1 can be determined as the average velocity from q^o to q^2.

This method of generating the trajectory as well as the method (4.4.11)-(4.4.15) provides for continuity of acceleration and the exact positioning in the user-supplied points.

Motion synthesis beside the procedures described above, can be performed optimally with respect to some performance criterion. Time suboptimal motion was examined in [47, 48]. This method assumes that the acceleration q(t) = 0 on medial parts of trajectory segments, while it takes a constant nonzero value during transitions from one trajectory segment to another. The optimization is based on linear programming technique.

Joint-interpolated motion is simpler and computationally less expensive then the Cartesian motion, since it requires inverse problem solution only for the trajectory segments' end points, and not along the whole trajectory. This feature is useful, especially in on-line motion generation. It is also used to move the manipulator from one configuration into another, when a kinematic degeneracy is located between them. On the other hand, it does not provide for straight-line motion of the end-effector in the space of the external coordinates. However, the joint-interpolated motion may be used as a very good approximation of the Cartesian path motion, if the interpolation is carried out between very close Cartesian points (several centimeters). In this situation, x-spline function, one kind of cubic spline which requires only local knot information for curve fitting, may be used [108].

4.4.2. External coordinates motion

In the cases when the tracking of strictly defined paths, in the manipulator work space is desired, motion should be specified as the continuous trajectory $x_e(t)$, $t \in [0, T]$ in the space of the external coordinates. This motion has also been referred to as orthogonal or Cartesian motion [1, 52]. It is also used when making accomodation motions, when sensors are employed, etc. Once the trajectory $x_e(t)$ is specified, the corresponding trajectory q(t) is evaluated by solving the inverse kinematic problem along the path.

Most frequently, motion generation is performed by specifying the set of points x_e^k, k=0,...,N, and iterpolating straight-line trajectories between these points

$$x_e(t) = x_e^k + \lambda(t)(x_e^{k+1}-x_e^k), \quad t_k < t < t_{k+1} \quad (4.4.20)$$

The parameter $\lambda(t) \in [0, 1]$ specifies the velocity distribution along the trajectory. Various types of velocity profiles (4.4.3) - (4.4.7) introduced for joint-interpolated motion, can also be applied to external coordinates motion synthesis. Thus, the functional requirements are fulfilled in a simple manner. However, in order to provide for continu-

ity of velocity and acceleration the motion is stopped at each user-
-specified point x_e^k.

In cases where exact positioning in the nostopping points is not required, the method (4.4.8) - (4.4.9) can be applied to the external
coordinates in the same way as for the joint coordinates. In the transition region, acceleration $\ddot{x}_e(t)$ varies according to a quadratic function between the zero values at the beginning and at the end of the interval, providing for continuity in acceleration. However, a position
error is introduced, which depends on the ratio between the transition
time τ_k and the execution time T_k of the entire segment.

A method for motion synthesis in the space of the external coordinates
suboptimally with respect to the total execution time, was proposed in
[49]. This method is analogous to the method [47], but constraints on
T_k and τ_k now become nonlinear, so that the linear programming technique cannot be applied. The problem is solved by the method of approximate programming.

4.6. Summary

Industrial manipulation tasks sometimes require motion along geometrically defined curves, such as circles, ellipses, parts of parabolas,
etc. Such curves can be described by their parameter equations. In order to facilitate specification of such paths, another Cartesian coordinate frame O'x'y'z' can be introduced, with respect to which the curve
will be simply described. The position and orientation of this frame
with respect to the base reference frame is defined by the position
vector $\vec{r}_c = [x_c \; y_c \; z_c]^T$ of the origin of the system O'x'y'z', and the
Euler angles ψ_c, θ_c, and φ_c. The relationship between coordinates x',
y' and z' and the coordinates x, y and z of the base system is given by

$$[x \; y \; z]^T = T(\psi_c, \theta_c, \varphi_c)[x' \; y' \; z']^T + \vec{r}_c \qquad (4.5.1)$$

where $T(\psi_c, \theta_c, \varphi_c) \in R^{3 \times 3}$ is the constant transformation matrix of the
type (1.4.3) (Chapter 1, Subsection 1.4.1). The curve representing the
manipulator hand's path is now specified with respect to this frame as

$$x_e' = f'(\lambda(t)), \qquad 0 \leq t \leq T \qquad (4.5.2)$$

where $x'_e = [x' \; y' \; z']^T$ is the position vector in system O'x'y'z', $\lambda(t) \in [\lambda^O, \lambda^F] \in R^n$ is the scalar motion parameter which affects the velocity distribution, and λ^O and λ^F correspond to the initial and final point on the trajectory. In this case the proportionality between velocities $\dot{\lambda}(t)$ and $\dot{x}_e(t)$ no longer exists.

This motion specification is suitable for describing plane curves. The plane O'x'y' can be chosen in such a way that the curve lies on it. Examples for such paths are circular motion, when Equation (4.5.2) becomes

$$x' = a \; \sin(\lambda(t))$$
$$y' = a \; \cos(\lambda(t)) \qquad\qquad (4.5.3)$$
$$z' = 0$$

or elliptic trajectory

$$x' = a \; \sin(\lambda(t))$$
$$y' = b \; \sin(\lambda(t)) \qquad\qquad (4.5.4)$$
$$z' = 0$$

The time history of the motion parameter $\lambda(t)$ should provide for smooth, jerkless motion.

In the case when a trajectory is strictly defined and very complex, tabular motion specification is used and the points specifying the task are very close to each other.

4.5. Procedurally Defined Motion

Methods for manipulator motion generation have been reviewed in this chapter. They provide for the continuity of position, velocity and acceleration, which are so desirable in robot motion. At the same time, the functionality of motion is achieved by attaining the positions and orientations required. The methods are simple and appropriate for practical application. They are based on manipulator kinematic equations only, and no dynamic characteristics of the system are taken into consideration.

Chapter 5
Dynamic Approach to Motion Generation

5.1. Introduction

Various methods for robot motion generation, based on the kinematic model of the manipulator, have been presented in the previous chapter. The problem of distributing motion to individual degrees of freedom is thus resolved. Kinematical constraints on maximal joint coordinates, velocity and acceleration are taken into account.

We will consider dynamic approach to manipulation system motion generation in this chapter, as opposed to this approach. The main feature of this approach is that the manipulator is considered as a dynamic system, modelled by dynamic equations of the mechanism and possibly by the model of the actuators in the joints. Dynamic motion synthesis is carried out optimally with respect to some dynamic performance index (e.g. energy, time), together with the constraints which are also dynamical in nature, such as constraints on maximal driving torques (forces) or maximal input signals of the actuators (input voltage for electric motors, input current of the servovalves for hydraulic actuators). The motion synthesis will include only the stage of nominal regimes, when no interferences are affecting the system, while the closed loop control problem will not be considered in this book (see [7, 58]).

In this chapter we will first give a short discription of the dynamic model of the manipulation system. This model includes model of the mechanical part of the system and models of actuators (we will consider electric DC motors and hydraulic actuators). Then, a short reviews of the methods of dynamic synthesis of manipulator trajectories proposed so far, will be given [59 - 64]. Afterwards, energy optimal motion generation will be analyzed, since energy optimal motion is very convenient for use in industry. Energy optimal motion also provides for minimal strains on the actuators and the mechanism itself. Time optimal motion however, is of the bang-bang type, resulting in abrupt variations of control signals and increased actuator strain. Energy-optimal motion assumes even greater significance in manipulation of heavy objects by means of powerful manipulators and in high speed motions with increased energy consumptions.

A method of off-line, energy optimal motion synthesis will be presented in Section 5.4. Given an end-effector trajectory in the space of the external coordinates, energy optimal velocity distribution is determined by using the dynamic programming algorithm. The method takes the complete dynamic model of the mechanism and the actuators into account, together with all the kinematic and dynamic constraints.

Real-time dynamic motion generation will be discussed in Section 5.5. A method of energy suboptimal trajectory synthesis in real time, based on a decentralized, simplified dynamic model of the system, will be presented.

5.2. Manipulation System Dynamic Model

The dynamic model of a manipulation system is comprised of the dynamic model of the mechanical system part and the model of the actuators driving the manipulator joints.

Dynamic equations of an n degree-of-freedom manipulator represent a set a of n nonlinear, second order differential equations. They describe the motion of the system, given the driving torques (forces) acting at manipulator joints. The manipulator is considered to be a set of rigid bodies interconnected by joints in the form of simple kinematic pair. Each link is characterized by its mass, tensor of inertia or three main central moments of inertia, in the case when the link coordinate frame is centered at the link mass center and its axes coincide with the main axes of inertia of the link.

Dynamic equations of a manipulator can be obtained by applying any of the methods known from classical mechanics: Lagrange's equations, Newton-Euler equations, Gibbs-Appel's equations [2-3, 5-6, 16-18, 65-68]. Independent of the modelling technique utilized, the relationship between the driving torque (force) P_i acting at joint i and the joint coordinates q, velocities $\dot{q}$ and accelerations $\ddot{q}$, is obtained in the form

$$P_i = \sum_{j=1}^{n} H_{ij}(q)\ddot{q}_j + \sum_{j=1}^{n} \sum_{k=1}^{n} C_{ijk}(q)\dot{q}_j\dot{q}_k + g_i(q), \quad i=1,\ldots,n \quad (5.2.1)$$

where P_i - is the driving torque (force) acting on joint i, $H_{ij}\ddot{q}_j$ - the inertial force acting on joint i as the result of the acceleration of joint j, $C_{ijk}\dot{q}_j\dot{q}_k$ - Coriolis forces acting on joint i as the result of the rotation of joints j and k, g_i - the gravitational force acting on joint i.

Equations (5.2.1) can be equivalently represented in the following matrix form

$$P = H(q)\ddot{q} + h(q, \dot{q}) \qquad\qquad (5.2.2)$$

where $q = [q_1 \cdots q_n]^T \epsilon Q \epsilon R^n$ is the vector of joint coordinates, $P = [P_1 \cdots P_n]^T \epsilon R^n$ is the vector of driving torques (forces), $H(q): R^n \to R^{n \times n}$ is the inertial matrix of the mechanism, $h(q, \dot{q}): R^n \times R^n \to R^n$ - is the vector of gravitational, Coriolis and centrifugal forces. Forces due to friction in the joints will be included in the actuators' models.

A large number of methods have been developed so far, aimed at facilitating and speeding the computation of the dynamic equations [16-18, 65-68]. Two investigation directions may be recognized: development of numeric, computer-oriented algorithms and development of symbolic modelling techniques. Due to the complexity of the dynamic equations, numeric computer-oriented methods had priority over symbolic ones. Symbolic dynamic models have been derived manually for simple manipulators, frequently in a simplified manner (neglecting Coriolis terms). Most recently, however, a new method has been developed [16-18], which makes possible the automatic, computer-assisted generation of the complete dynamic model of an arbitrary serial-link manipulator, in numeric-symbolic form. The parameters are treated as numerical constants, while the joint coordinates and velocities are considered as symbolic variables. As compared with the numerical methods, this algorithm reduces the model evaluation period by the order of magnitude [16, 69].

We shall not analyze any of the methods for dynamic modelling in this book, since the method of computing the model is not of great importance for our optimal motion synthesis. Obviously, the faster the model computation, the faster the evaluation of the optimal trajectory.

Electric DC motor model

One of the actuators which is most frequently utilized for driving industrial robots is certainly the electric DC motor. The equations describing its operation are as follows

$$u(t) = L \frac{di}{dt} + Ri(t) + K_{me}N_v\dot{q}(t)$$

$$P(t) = K_{em}N_m i(t) - J\ddot{q}(t) - F\dot{q}(t)$$

(5.2.3)

where L is the rotor coil inductance; R - the rotor resistance; J - the moment of rotor inertia reduced to the output shaft $J = J_m N_m N_v$, F - the viscous friction coefficient reduced to the output shaft $F = F_m N_v N_m$; N_v - the angular velocity reduction ratio from the input to the output shaft of the reducer, N_m - the torque reduction ratio from the output to the input shaft of the reducer; K_{me} - the mechanical-electrical constant of the motor; K_{em} - the electro-mechanical constant of the motor, i - the rotor current, $\dot{q}$ - the angular velocity of the output motor shaft (upon the reduction), $\ddot{q}$ - the angular acceleration of the output shaft, u - the rotor input voltage, P - the driving torque attained by the motor on its output shaft.

The model (5.2.3) can be equivalently represented in the following matrix form

$$\dot{x} = Ax + bu + fP$$

(5.2.4)

where $x = [q \ \dot{q} \ i]^T$ is the state vector, $A \in R^{3 \times 3}$, $b \in R^3$, $f \in R^3$ are given by

$$A = \begin{bmatrix} 0 & 1 & 0 \\ 0 & -\frac{F}{J} & \frac{K_{em}N_m}{J} \\ 0 & \frac{-K_{me}N_v}{L} & \frac{-R}{L} \end{bmatrix}$$

(5.2.5)

$$b = \begin{bmatrix} 0 \\ 0 \\ \frac{1}{L} \end{bmatrix}, \qquad f = \begin{bmatrix} 0 \\ -\frac{1}{J} \\ 0 \end{bmatrix}$$

(5.2.6)

Model (5.2.4) is a linear, stationary model of the third order. The control signal u(t) is subject to the amplitude constraint

$$|u| \leqslant u_{max}$$

(5.2.7)

where u_{max} is the maximal input voltage of the motor.

In the case when the rotor inductance is low, the third order model (5.2.4) can be reduced to a linear system of the second order. It can also be represented in form (5.2.4), with $x = [q \; \dot{q}]^T$ and

$$A = \begin{bmatrix} 0 & 1 \\ 0 & -\dfrac{F}{J} - \dfrac{K_{em}K_{me}N_mN_v}{RJ} \end{bmatrix} \qquad (5.2.8)$$

$$b = \begin{bmatrix} 0 \\ \dfrac{K_{em}N_m}{RJ} \end{bmatrix}, \qquad f = \begin{bmatrix} 0 \\ -\dfrac{1}{J} \end{bmatrix} \qquad (5.2.9)$$

Hydraulic actuators models

Another type of actuators, frequently encountered in practice, is a hydraulic cylinder controlled by a hydraulic servovalve. These actuators are modelled by a set of nonlinear differential equations of the fifth order [70]. However, with servovalves with a wide bandwidth, the third order linearized model can be adopted [71]. It can be described by the following set of equations

$$Q = S_c\dot{\ell} + C_{t\ell}p + \frac{V_m}{4\beta}\dot{p}$$

$$S_c p = m_t\ddot{\ell} + B_c\dot{\ell} + F_t \qquad (5.2.10)$$

$$Q = K_q i - K_c p$$

where Q is the fluid load flow $[m^3/s]$, p - the difference between the preassures on both sides of the piston (load preassure) $[N/m^2]$, ℓ - the linear velocity of the piston $[m/s]$, $\ddot{\ell}$ - the linear acceleration of the piston $[m/s^2]$, S_c - the piston area $[m^2]$, $C_{t\ell}=(C_{i\ell}+C_{e\ell})/2$, $C_{i\ell}$ and $C_{e\ell}$ are internal and external oil leakage coefficients, respectively $[m^3/s/N/m^2]$, V_m - the cylinder (working) volume $[m^3]$, β - the oil compressibility coefficient $[N/m^2]$, m_t - the piston mass (including the rod mass) [kg], B_c - the viscous friction coefficient [kgm/s], F_t - the external force acting on the piston [N], K_c - the gradient of the servovalve flow--preassure characteristic $[m^3/s/N/m^2]$, K_q - the flow-current coefficient

$[m^3/s/mA]$.

The Equations (5.2.10) can be presented in the form of a state space model

$$\dot{x} = Ax + bu + fP$$

where the state vector $x \in R^3$, $x = [\ell \ \dot{\ell} \ p]^T$

$$A = \begin{bmatrix} 0 & 1 & 0 \\ 0 & -\dfrac{B_c}{m_t} & \dfrac{S_c}{m_t} \\ 0 & -\dfrac{S_c 4\beta}{V_m} & -\dfrac{(K_c + C_{t\ell})4\beta}{V_m} \end{bmatrix}$$

(5.2.11)

$$b = \begin{bmatrix} 0 \\ 0 \\ \dfrac{K_q 4\beta}{V_m} \end{bmatrix}, \qquad f = \begin{bmatrix} 0 \\ -\dfrac{1}{m_t} \\ 0 \end{bmatrix}$$

Here, the control is the servovalve current i, while the generalized torque P is the force exerted on the external load by the piston. The model of the vane hydraulic actuators, having rotational motion, can be derived analogously.

Dynamic model of a complete manipulation system

The complete dynamic model of a manipulation system is obtained by joining model (5.2.2) of the mechanical system part to the actuators' models. The model of the actuator i, driving the joint i, $i=1,\ldots,n$, will be adopted in the form of the linear stationary system

$$S^i: \quad \dot{x}^i = A^i x^i + b^i N(u^i) + f^i P_i, \qquad i=1,\ldots,n \qquad (5.2.12)$$

where $x^i \in R^{n_i}$ is the state vector of the model of actuator i, n_i is the order of the model, $u^i \in R^1$ is the control input of actuator i, $P_i \in R^1$ is the actuator output torque (force), $A^i \in R^{n_i \times n_i}$ is the matrix of sybsystem S^i, $b^i \in R^{n_i}$ is the input distribution vector, $f^i \in R^{n_i}$ is the load distribution vector, $N(u^i)$ is the nonlinearity of the amplitude saturation type

$$N(u^i) = \begin{cases} -u^i_{max}, & u^i < -u^i_{max} \\ u^i & -u^i_{max} < u^i < u^i_{max}, \\ u^i_{max} & u^i > u^i_{max} \end{cases} \qquad i=1,\ldots,n \qquad (5.2.13)$$

where $-u^i_{max}$ and u^i_{max} are the boundaries of linear actuator operation.

The integral model of all the actuators is obtained by uniting the models S^i

$$S^A: \quad \dot{x} = Ax + BN(u) + FP \qquad (5.2.14)$$

where $x = [q^T \; \dot{q}^T \; x_r^T]^T \in R^N$ is the state vector of the complete system comprising the state vector of the mechanical system

$$\xi = [q^T \; \dot{q}^T]^T, \qquad \xi \in R^{2n} \qquad (5.2.15)$$

and the vector $x_r \in R^{N-2n}$, $N = \sum_{i=1}^{n} n_i$. This vector x_r consists of the remaining state coordinates (motor currents, load preassures for hydraulic cylinders). The matrices $A \in R^{N \times N}$, $B \in R^{N \times n}$, $F \in R^{N \times n}$ are simply obtained once the matrices A^i, b^i and f^i, $i=1,\ldots,n$ are known; $N(u) = [N(u^1) \cdots N(u^n)]^T$ and $P = [P_1 \cdots P_n]^T$ denote the control vector and the driving torque vector, respectively.

In order to form the state space model of the complete system, we will present the manipulator dynamic equations

$$P = H(q)\ddot{q} + h(q, \dot{q}) \qquad (5.2.16)$$

in the form

$$P = H_m(\xi)\dot{\xi} + h(\xi) \qquad (5.2.17)$$

where $H_m(\xi) = [0 \mid H(q)] \in R^{n \times 2n}$, $\xi \in R^{2n}$ is given by (5.2.15). By eliminating the driving torque vector from Equations (5.2.14) and (5.2.17), we obtain

$$\dot{x} = Ax + BN(u) + FH_m(\xi)\dot{\xi} + Fh(\xi) \qquad (5.2.18)$$

By introducing

$$F_m = [FH_m(\xi) \mid 0] \in R^{N \times N} \qquad (5.2.19)$$

the model (5.2.18) becomes

$$\dot{x} = Ax + BN(u) + F_m\dot{x} + Fh \qquad (5.2.20)$$

where $x = [\xi^T \ x_r^T]^T = [q^T \ \dot{q}^T \ x_r^T]^T \in R^N$. The model of the complete manipulation system S is now obtained from (5.2.20) in the following form

$$S: \ \dot{x} = \hat{A}(x) + \hat{B}(x)N(u) \qquad (5.2.21)$$

where $\hat{A}(x): \ R^N \rightarrow R^N$, $\hat{B}(x): \ R^N \rightarrow R^{N \times n}$ are the system and gain matrices respectively given by

$$\hat{A}(x) = (I_n - F_m(x))^{-1}(Ax + Fh(x))$$

$$\hat{B}(x) = (I_n - F_m(x))^{-1}B \qquad (5.2.22)$$

Obviously, the matrices of the complete model are complex, nonlinear
functions of the state vector and they depend on the mechanism and actuators' parameters.

Let us notice here that the dynamic model of the mechanism (5.2.17)
(without the actuators) can also be presented in the canonic form

$$\dot{\xi} = A_m(\xi) + B_m(\xi)P \qquad (5.2.23)$$

where the matrices $A_m(\xi): \ R^{2n} \rightarrow R^{2n}$ and $B_m(\xi): \ R^{2n} \rightarrow R^{2n \times n}$ are simply obtained once the matrice $H(q)$ and $h(q, \dot{q})$ are known.

Nominal trajectory

A state space trajectory $x(t) \in R^N$ for the manipulation system modelled
by (5.2.21), corresponding to a given manipulation task, will be referred to as a nominal trajectory. Once the task is specified and the
time histories $q(t)$, $\dot{q}(t)$, $\ddot{q}(t)$ are evaluated, the nominal trajectory
$x(t)$ is completely defined. Then, the controls signals are given by

$$u(t) = (\hat{B}^T\hat{B})^{-1}\hat{B}^T(\dot{x}-\hat{A}(x))$$

(5.2.24)

From the numerical point of view it is more suitable to evaluate the driving torques $P(q, \dot{q}, \ddot{q})$ first, and then substitute them into the actuator models in order to obtain the control signals $u(t)$.

5.3. An Overview of Methods for Dynamic Motion Synthesis

As we observed in the previous section, the evaluation of the nominal trajectories includes the evaluation of the state space trajectories $x(t) \in X \subset R^n$ and the corresponding nominal control $u(t) \in R^n$ for the system modelled by

$$S: \quad \dot{x} = \hat{A}(x) + \hat{B}(x)N(u)$$

(5.3.1)

where $x = [q^T\ \dot{q}^T\ x_r^T]^T \in R^N$ is the state vector of the complete system, $\hat{A}(x) \in R^N$ and $\hat{B}(x) \in R^{N \times n}$ are nonlinear dynamic model matrices, $N(u)$ - the control vector subjected to amplitude constraints.

As with kinematic motion synthesis, the trajectory $x(t)$ should answer to certain functional demands (e.g. provides for straight line motion), and if possible, minimize a performance index, also taking into account the constraints on input signals or driving torques (forces). The same holds for the case when only the dynamic model of the mechanical system (5.2.23) is considered, with the state vector being $x = \xi = [q^T\ \dot{q}^T]^T$.

The dynamic system (5.3.1) represents a continuous, nonlinear, stationary, multivariable system (we will assume that the parameters are not time varying). The evaluation of the elements of matrices $\hat{A}(x)$ and $\hat{B}(x)$ is carried out automatically, by means of a computer, since they are very complex, nonlinear functions of the state vector. In the following text we will analyze the problems arising in optimal nominal trajectory synthesis by solving the canonical system of equations.

Let us assume for the moment that the control $u(t)$ is not subject to the amplitude constraints. The functions $\hat{A}(x)$ and $\hat{B}(x)$ are continuous, differentiable functions of x, as will be shown later. We will also assume that the terminal conditions for a part of the state coordinates are also imposed

$$x_i(t_1) = x_i^F, \qquad i=1,\ldots,i_1$$

where t_1 is the terminal time instant, while the initial state is completely defined $x(t_o) = x^o$.

In the general case, the optimal nominal trajectory synthesis is obtained by minimizing the cost functional [72]

$$J(u, x^o) = \int_{t_o}^{t_1} L(x, u)dt + g[x(t_1)] =$$

$$= \int_{t_o}^{t_1} L(x, u)dt + g[x_{i_1+1}(t_1), \ldots, x_N(t_1)] \qquad (5.3.2)$$

where $L(x, u)$ is a scalar, continuous, differentiable function of the state vector and the control, t_1-t_o is the fixed motion execution time, g - a scalar-valued function of the remaining nonspecified state coordinates in the terminal time instant.

The optimal control problem is to determine the control $u^*(t)$, $t \in [t_o, t_1]$ which transfers the system from the state $x(t_o) = x^o$ to the state $x_i(t_1) = x_i^F$, $i=1, \ldots, i_1$ and minimizes the chosen performance functional $J(u, x^o)$ over all such controls.

The optimal control $u^*(t)$ and the corresponding optimal trajectory $x^*(t)$ and the costate vector $p^*(t)$ are the solution of the canonical system [72, 73]

$$\dot{x}^*(t) = \frac{\partial H}{\partial p}^T(x^*(t), p^*(t), u^*(t)), \quad x^*(t_o) = x^o, \quad x_1^*(t_1) = x_1^F,$$

$$i=1,2,\ldots,i_1 \qquad (5.3.3)$$

$$\dot{p}^*(t) = -\frac{\partial H}{\partial x}^T(x^*(t), p^*(t), u^*(t)), \quad p_i^*(t_1) = \frac{\partial g^T(x(t_1))}{\partial x_i},$$

$$i=i_1+1,\ldots,N \qquad (5.3.4)$$

$$\frac{\partial H}{\partial u}^T(x^*(t), p^*(t), u^*(t)) = 0 \qquad (5.3.5)$$

where $H(x, p, u)$ is the Hamiltonian given by

$$H(x, p, u) = L(x, u) + p^T(\hat{A}(x) + \hat{B}(x)u) \qquad (5.3.6)$$

The order of the above system is relatively high and equals 2N+n. For a 6 degree-of-freedom manipulator and the actuators modelled by the third order models, this amounts to 42.

If the quadratic performance index is chosen

$$J(u, x^o) = \int_{t_o}^{t_1} (\frac{1}{2} u^T Ru + x^T Su)\,dt$$

where R is symmetric positive definite matrix, and S is a symbolic positive semidefinite matrix, Equations (5.3.3) - (5.3.5) reduce to

$$\dot{x}^*(t) = \hat{A}(x^*) + \hat{B}(x^*)u^*(t), \quad x^*(t_o) = x^o, \quad x_i^*(t_1) = x_i^F,$$

$$i=1,\ldots,i_1$$

$$\dot{p}^*(t) = -Su^*(t) - (\frac{\partial \hat{A}^T(x^*)}{\partial x} + u^{*T}(t) \frac{\partial \hat{B}^T(x^*)}{\partial x})p^*(t), \quad p_i^*(t_1) = 0,$$

$$i=i_1+1,\ldots,N \quad (5.3.7)$$

$$Ru^*(t) + S^T x^*(t) + \hat{B}^T(x^*)p^*(t) = 0$$

The solution of the optimal trajectory requires the evaluation of the nonlinear functions $\hat{A}(x)$ and $\hat{B}(x)$ at each integration interval, together with the evaluation of the partial derivatives of these functions, i.e. the matrices $\frac{\partial \hat{A}(x)}{\partial x}$ and $\frac{\partial \hat{B}(x)}{\partial x}$.

They can be determined thus: in Section 5.2 the dynamic model matrices have been derived in the form

$$\hat{A}(x) = (I_N - F_m(x))^{-1}(Ax + Fh(x))$$

$$(5.3.8)$$

$$\hat{B}(x) = (I_N - F_m(x))^{-1}B$$

where $A \in R^{N \times N}$, $B \in R^{N \times n}$, $F \in R^{N \times n}$ are the constant matrices of the actuator model (5.2.14), $h(x): R^N \to R^N$ - the vector of the mechanism model (5.2.16), $F_m(x): R^N \to R^{N \times N}$ is the matrix given by (5.2.19).

By differentiating Equations (5.3.8) we obtain

$$\frac{\partial \hat{A}(x)}{\partial x} = (I_N - F_m(x))^{-1}[\frac{\partial F_m(x)}{\partial x}(I_N - F_m(x))^{-1}(Ax + Fh(x)) + A + F\frac{\partial h(x)}{\partial x}]$$

$$\frac{\partial \hat{B}(x)}{\partial x} = (I_N - F_m(x))^{-1}\frac{\partial F_m(x)}{\partial x}(I_N - F_m(x))^{-1}B$$

According to (5.2.19) the partial derivative $\frac{\partial F_m(x)}{\partial x}$ is given by

$$\frac{\partial F_m(x)}{\partial x} = [F \frac{\partial H_m(x)}{\partial x} \mid 0]$$

where

$$\frac{\partial H_m(x)}{\partial x} = [0 \mid \frac{\partial H}{\partial x}]$$

The inertia matrix $H(q): R^n \rightarrow R^{n \times n}$ does not depend on the complete state vector $x = [q^T \ \dot{q}^T \ x_r^T]^T$, but only on joint coordinates q. Thus we can write

$$\frac{\partial H}{\partial x} = \alpha_{ij}^k, \qquad i,j=1,\ldots,n, \qquad k=1,\ldots,N$$

where

$$\alpha_{ij}^k = \begin{cases} a_{ij}^k & i,j,k=1,\ldots,n \\ \\ 0 & i,j=1,\ldots,n, \qquad k=n+1,\ldots,N \end{cases}$$

with a_{ij}^k, $i,j=1,\ldots,n$ being the elements of matrix $H_p \in R^{n \times n \times n}$

$$H_p = \frac{\partial H}{\partial q} = [a_{ij}^k]$$

The partial derivative $\frac{\partial h}{\partial x}$ can be written in the form

$$\frac{\partial h}{\partial x} = [\frac{\partial h}{\partial q} \ \frac{\partial h}{\partial \dot{q}} \ \frac{\partial h}{\partial x_r}]$$

or

$$\frac{\partial h}{\partial x} = [h_p \ h_v \ 0]$$

where the matrice $h_p \in R^{n \times n}$ and $h_v \in R^{n \times n}$ are given by

$$h_p = \frac{\partial h}{\partial q}, \qquad h_v = \frac{\partial h}{\partial \dot{q}}$$

Thus, the evaluation of the partial derivatives $\frac{\partial \hat{A}(x)}{\partial x}$ and $\frac{\partial \hat{B}(x)}{\partial x}$ has been reduced to the evaluation of the matrices H_p, h_p and h_v which re-present the matrices of the linearized manipulator dynamic model. The evaluation of these matrices is a very complex procedure, so that only computer-aided methods are applicable. A numerical method for evalua-ting the linearized dynamic model (without the use of numerical dif-ferentiation) was developed in [74]. Most recently, however, a numeric--symbolic method for automatic, computer aided linearized dynamic model generation was developed [16], which is a logical continuation of the method for symbolic dynamic modelling of serial-link manipulators.

The system of equations (5.3.7) can be solved in two ways using the first-order gradient procedures (the algorithms based on Hessian matrices cannot be applied in this case, because of the complexity of the model matrices). The first method is based on eliminating the control vector from the first two equations of the system (5.3.7), using the third equation

$$u^*(t) = -R^{-1}(S^T x^*(t) + \hat{B}^T(x^*)p^*(t))$$

$$\dot{x}^*(t) = \hat{A}(x^*) - \hat{B}(x^*)R^{-1}(S^T x^*(t) + \hat{B}^T(x^*)p^*(t)),$$

$$x^*(t_o) = x^o$$

$$x_i^*(t_1) = x_i^F, \quad i=1,\ldots,i_1 \qquad (5.3.9)$$

$$\dot{p}^*(t) = SR^{-1}S^T x^*(t) + x^{*T}(t)SR^{-1}\frac{\partial \hat{B}(x^*)}{\partial x}p^*(t) +$$

$$+ (SR^{-1}\hat{B}^T(x^*) - \frac{\partial \hat{A}(x^*)}{\partial x})p^*(t) + p^{*T}(t)\hat{B}(x^*)R^{-1}\frac{\partial \hat{B}(x^*)}{\partial x}p^*(t),$$

$$p_i^*(t_1) = 0, \quad i = i_1+1,\ldots,N$$

The solution of the above system of 2N equations is not easily obtainable, because the initial costate $p(t_o)$ is not known. The system can be solved by supposing an initial value for $p(t_o)$ and by integrating the system (5.3.9), given the initial $x(t_o) = x^o$. Once the solution is obtained, the extent should be checked to which the boundary conditions in the terminal time instant have been satisfied, i.e. to evaluate the error vector $e(t_1)$ as

$$e(t_1) = \begin{cases} x_i(t_1) - x_i^F, & i=1,\ldots,i_1 \\ p_i(t_1) & i=i_1+1,\ldots,N \end{cases}$$

If an error appears, a correction Δp_o should be found to reduce the error. This procedure should be repeated until the error is reduced to a predetermined value ε. This procedure, however, is difficult to realize, since the costate p may behave unstably, complicating the integration.

The second approach to solving the system (5.3.7) is an algorithm in the control space. It is based on an exact solution of the first two equations, while the error in the third is reduced by an iterative procedure. Let us suppose some initial control $u_o(t)$, $t\in[t_o, t_1]$. The corresponding nominal trajectory $x_o(t)$ is evaluated from the first equation of the system (5.3.7)

$$\dot{x}_o(t) = \hat{A}(x_o) + \hat{B}(x_o)u_o(t)$$

The control $u_o(t)$ supposed should provide that $x_{io}(t_1) = x_i^F$, $i=1,\ldots,i_1$. This can easily be achieved (e.g. the control which transfers the system from the initial to the final state along a straight line). Now, the second equation of the system (5.3.7) should be solved, for the trajectory $x_o(t)$, $u_o(t)$ determined. There are $N-i_1$ boundary conditions. The remaining i_1 boundary conditions for $p_{io}(t_1)$, $i=i_1,\ldots,i_1$ should be assumed, and then the second equation can be solved starting from the terminal time internal t_1. Finally, the extent to which the third equation of the system (5.3.7) is satisfied, must be checked, and the correction for the control introduced

$$\delta u_o(t) = -k \frac{\partial H^T}{\partial u}(x_o(t), p_o(t), u_o(t)) =$$

$$= -k(Ru_o(t)+S^Tx_o(t)+\hat{B}^T(x_o)p_o(t))$$

where k is a positive constant. In the next iteration the control is taken to be

$$u_1(t) = u_o(t) + \delta u_o(t)$$

The question arises, however, whether the control $u_1(t)$ still satisfies the terminal boundary conditions. If it doesn't, the initial guess for the costate $p_{io}(t_1)$, $i=1,\ldots,i_1$ and the step size k should be changed. If $u_1(t)$ transfers the system to the terminal state, the minimization process continues until the inequality

$$\int_{t_o}^{t_1} ||\frac{\partial H^T}{\partial u}(t)||dt < \varepsilon$$

is satisfied, where ε is a given positive constant.

The above algorithm causes no problems during the equation integration. However, significant problems are encountered in choosing the values $p_i(t_1)$, $i=1,\ldots,i_1$, so that the terminal boundary conditions on the state vector are satisfied.

We can see from the above analysis that making an optimal nominal trajectory synthesis for a manipulator moving between two given points is a very complex and tedious task. Besides, this method does not consider the control constraints of the real manipulation system, nor the constraints on the trajectory itself. On the other hand manipulation tasks are well defined (e.g. straight line motion, given the disposition of obstacles to be avoided, etc.). Therefore it would be very difficult for the optimal solution to satisfy the functionality requirements even if the solution to the canonical system of equation was easily obtainable.

For these reasons, the methods for dynamic motion synthesis proposed so far, either make use of some other optimization techniques [62, 64], or simplified dynamic models [60-61]. We will here present a brief review of these methods.

The time-optimal motion synthesis

The time-optimal control of open-loop articulated kinematic chains, by solving the canonical system, was analyzed in [59]. The 3 degree-of--freedom manipulator shown in Fig. 5.1 was considered. The system is modelled by the symbolic dynamic model of the mechanism (5.2.23) (without the actuators), which is the equivalent to

$$\dot{x} = \hat{A}(x) + \hat{B}(x)P \qquad (5.3.10)$$

where $x = [q_1 \ q_2 \ q_3 \ \dot{q}_1 \ \dot{q}_2 \ \dot{q}_3]^T$ is the state vector, $P = [P_1 \ P_2 \ P_3]^T$ - the control vector representing the driving torques, $\hat{A}(x) \in R^6$, $\hat{B}(x) \in R^{6 \times 3}$ are the system matrices. The model (5.3.10) represents a set of 6 coupled, nonlinear, first order differential equations.

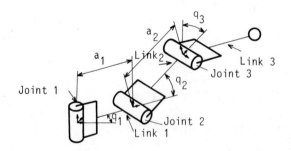

Fig. 5.1. The three degree-of-freedom manipulator

At the initial time $t=t_o$, the system is assumed to be in the state $x(t_o) = x^o$, while at the final time $t = t_1$ the state is required to satisfy the terminal constraint $x(t_1) = x^F$. The driving torques P_i are constrained in amplitude

$$|P_i(t)| \leqslant P_{imax} \qquad \forall t \in [t_o, t_1], \ i=1,2,3 \qquad (5.3.11)$$

where P_{imax} are positive constants.

The solution to the time-optimal control problem is given by the admissible control which minimizes the cost functional

$$J = \int_{t_o}^{t_1} 1 \ dt = t_1 - t_o \qquad (5.3.12)$$

The search for the optimal solution is considerably narrowed by the minimum principle of Pontryagin [72]. In this case the Hamiltonian (5.3.6) becomes

$$H(x, p, P) = p^T(\hat{A}(x) + \hat{B}(x)P) + 1 \qquad (5.3.13)$$

where $p \in R^6$ is the costate vector (adjoint vector). The optimal solution should satisfy the canonical system of equations (5.3.3)-(5.3.4).

Now let $P^*(t)$ be an optimal control and let $x^*(t)$ denote the optimal solution of equations (5.3.10), which is the result of the application of this control starting from x^o. Then the minimum principle is stated as follows: if $P^*(t)$ is the optimal control and $x^*(t)$ the corresponding optimal trajectory, then there exists an adjoint vector $p^*(t)$, satisfying (5.3.4), so that, at every time instant t, $t_o \leqslant t \leqslant t_1$

$$H(x^*, p^*, P^*) \leqslant H(x^*, p^*, P) \qquad (5.3.14)$$

for all admissible controls P. The optimal control is, therefore, the control which minimizes the Hamiltonian at every instant within the given time interval.

By substituting the analytical expressions for the matrices $\hat{A}(x)$, $\hat{B}(x)$ for the manipulator shown in Fig. 5.1, into (5.3.13), the Hamiltonian is obtained as an explicit function of the state vector x, costate p and the driving torques P. Using the minimum priniple and by inspection of the Hamiltonian, the optimal control signals are obtained in the form

$$P_i^* = -P_{imax} \text{sgn} \ f_i(x, p), \qquad i=1,2,3 \qquad (5.3.15)$$

where $f_i(x, p)$ are nonlinear scalar functions of the state vector x and the costate p.

The Equation (5.3.15) shows that the optimal control is bang-bang and the sign reversals depend nonlinearly on both the state and the adjoint variables. If these equations are used to eliminate P_1, P_2 and P_3 from the canonical system of equations (5.3.3) - (5.3.4), the result is a set of 12 first-order nonlinear differential equations in the variables x_i and p_i, i=1,...,6. The optimization problem is then reduced to a two-point boundary value problem. Owing to the nonlinearity of the equations involved, a numerical approach is necessary for the solution of this problem. In [59] an example of the optimal joint trajectories and the corresponding optimal control signals is given. Bearing in mind the fact that the control is nominal, i.e. it does not take any unexpected disturbances which may act on the system into account, the optimal motion synthesis is too complex.

The time-optimal motion synthesis for the three degree-of-freedom manipulator shown in Fig. 5.2, has been considered in [60]. The joint axes are parallel which results in the plane manipulator motion. The first two links have the same length L, and negligible masses ($m_1=m_2=0$). The length of the third link equals zero, the mass m_3 and the moment of inertia of the third link are specified. The Cartesian coordinates x, y and the angle φ_3 (Fig. 5.2), specifying the gripper orientation, are considered to be external coordinates.

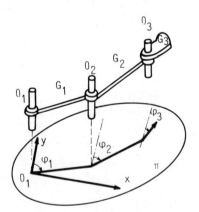

Fig. 5.2. The plane manipulator

The assumptions introduced, concerning the manipulator structure, make the derivation of the dynamic equations of motion, which relate between

the driving torques P_i and the external coordinates possible. The time-
-optimal control problem is to determine such admissible control $P_i^*(t)$,
i=1,2,3 which transfers the system from the initial position x^o, y^o to
the final position x^1, y^1 along a straight line trajectory. Optimization is carried out only for the first two degrees of freedom, while
the third joint is assumed to have $P_3(t) \equiv 0$, $t \in [t_o, t_1]$. The driving
torques are constrained in amplitude, $|P_i(t)| \leqslant P_{imax}$, i=1,2,3. The op-
timal control is obtained in the form

$$P_i(t) = P_{imax} \text{sgn} \, f(x, y), \qquad i=1,2$$

As we have seen, in this case the functionality requirements are satisfied (straight line motion is achieved), but the manipulator taken into
consideration is very simple, having practically two degrees of freedom
with a mass concentrated at the end of the second link.

The time-optimal control of mechanical manipulators was also analyzed
in [63]. The system is modelled by the dynamic model of the mechanism
without the actuators. The control algorithm determines the manner in
which a six degree-of-freedom manipulator should move along a specified
path in three dimensional space to complete its motion in minimum time
without violating actuator constraints (maximal torques). The time instants the switching from maximal acceleration to maximal deceleration
should be carried out, are determined in the process of finding the intersection points between the velocity limit curve and the curve obtained by integrating maximal acceleration or deceleration in time. The
algorithm significantly reduces the motion execution time. However, as
with the other time-optimal algorithms, the total execution time is
very sensitive to the time instants at which the switching occurs, i.e.
even small deviations from the optimal points would result in a considerable increase of the execution time. The locus of the switching points
is also very sensitive to the system's parameter variations.

Energy optimal motion synthesis

As we discussed in the introduction, energy optimal motion is more suitable for practical applications, since it yields smooth motion and minimizes the actuators' and mechanism strains together with energy consumption.

An interesting method for overcomming the complexity of optimal motion synthesis was proposed in [62]. By this method the problem of optimal motion synthesis is reduced to the problem of parameter optimization resolved by the dual simplex method. Namely, it is assumed that the optimal joint trajectory can be presented in the following form

$$q_i(t) = \psi_{io}(t) + \sum_{k=1}^{N} a_{ik}\psi_k(t), \qquad 0 < t < 1, \quad i=1,\ldots,n \qquad (5.3.16)$$

where the functions $\psi_{io}(t)$, ψ_k, $k=1,\ldots,N$ satisfy the boundary conditions

$$\psi_{io}(0) = q_i^O, \qquad \psi_{io}(1) = q_i^F$$

$$\dot{\psi}_{io}(0) = 0, \qquad \dot{\psi}_{io}(1) = 0 \qquad (5.3.17)$$

$$\psi_k(0) = \psi_k(1) = \dot{\psi}_k(0) = \dot{\psi}_k(1) = 0, \qquad k=1,\ldots,N$$

Without the loss of generality, it is assumed here that $t \in [0, 1]$, i.e. the time is normalized. In accordance with (5.3.17)

$$\psi_{io}(t) = q_i^O + (q_i^F - q_i^O)(3t^2 - 2t^3)$$

$$\psi_k(t) = (t^2 - 2t^3 + t^4)t^{k-1} \qquad (5.3.18)$$

may be chosen. The optimal trajectory is obtained as follows: by substituting the joint trajectory (5.3.16) into the symbolic dynamic equations of a particular mechanism, the corresponding driving torques are obtained as functions of the parameters a_{ik}. The optimal control problem is now reduced to the problem of parameter optimization

$$\min_{u \in U} J(u) \to \min_{a_{ik}} J(a_{ik})$$

where the cost functional J is adopted to be

$$J = \int_0^1 (u^T R u + x^T Q x) dt$$

Parameter optimization is carried out by the dual simplex method, taking all types of constraints on the state vector and the control into account.

The energy optimal motion for a 3 degree-of-freedom arthropoid manipulator, travelling along an elliptic path, was analyzed in [64]. The computational technique employed is to express the optimal velocity profile as a finite series of the sinus and cosinus type.

5.4. Determination of the Energy Optimal Velocity Distribution Using Dynamic Programming

As we observed in the previous section, the application of the classical optimal control theory to manipulator motion synthesis, is impeded by the complexity and nonlinearity of the dynamic equations. Besides, the optimal solution need not satisfy the functionality requirements since the constraints on the manipulator end-effector trajectory cannot easily be introduced.

An optimization method which can cope with such a highly nonlinear, complex system subjected to strong constraints, is the dynamic programming method. In this section, we will present an algorithm for optimal nominal trajectory synthesis, where the manipulation system is modelled by the complete dynamic model of the mechanisms and actuators [75-77].

This algorithm is based on the application of the dynamic programming method [78]. End-effector velocity distribution is optimized, given an end-effector trajectory in the external coordinates space and a desired cost functional.

Problem statement

Let us consider the manipulation system with n degrees of freedom, described by the continuous, nonlinear stationary system

$$\dot{x} = \hat{A}(x) + \hat{B}(x)N(u), \quad x(t_o) = x^o, \quad x(t_1) = x^F \qquad (5.4.1)$$

where $x = [q^T \ \dot{q}^T] \epsilon R^{2n}$ is the state vector composed of the joint coordinates and velocities, $\hat{A}(x): R^{2n} \to R^{2n}$, $\hat{B}(x): R^{2n} \to R^{2n \times n} \to$ the system matrices, $u \epsilon U \subset R^n$ is the admissible control constrained in amplitude (maximal input voltage for DC motors); x^o and x^F are the given initial and final system states, respectively, $t_1 - t_o$ - is the motion execution time.

Let us assume that the cost functional has the general form

$$J = \int_{t_o}^{t_1} L_1(x, u)dt + g_1[x(t_1)] \qquad (5.4.2)$$

where L_1 and g_1 are scalar-valued functions. We will also assume that the state and control vectors are subject to the constraints

$$x \in X \subset R^{2n}, \quad x(t_o) = x^o, \quad x(t_1) = x^F, \quad u \in U \subset R^n \qquad (5.4.3)$$

The admissible optimal control $u^*(t)$ which transfers the system from the initial state x^o along the optimal trajectory $x^*(t)$ into the terminal state x^F, minimizing the performance index (5.4.2), should be determined.

The problem solution

In order to solve the stated problem by using the dynamic programming method, we will consider the discretized form of the system model

$$x(k+1) = x(k) + \Delta t \hat{A}(x(k)) + \hat{B}(x(k)) u(k)$$

$$x(k) = x(k\Delta t + t_o),$$

$$u(k) = u(k\Delta t + t_o) \qquad (5.4.4)$$

$$\Delta t = \frac{t_1 - t_o}{N}$$

where N is the number of discretization intervals.

The cost functional (5.4.2) in the discrete domain becomes

$$J_o = \sum_{k=0}^{N-1} L(x(k), u(k)) + g[x(N)] \qquad (5.4.5)$$

where $L(x(k), u(k)) = \Delta t L_1(x(k), u(k))$ and $g(x(N)) = g_1(x(t_1))$.

The problem is in determining such a set of the control vectors $u^*(0), \ldots, u^*(N-1)$ and the corresponding state vectors $x^*(0), \ldots, x^*(N-1)$, as to minimize the cost functional J_o under the constraints (5.4.3).

Dynamic programming reduces this problem to a series of successive minimizations with respect to $u(k)$, $k=0, \ldots, N-1$, making use of the principle that the second part of the optimal trajectory is the optimal trajectory.

At the first optimization step we will consider the last time interval

[N-1, N]. For the various state vector values assumed, $x^m(N-1)\in X$, m=1,...
...,M^2 the corresponding controls $u^m(N-1)\in U$, which transfer the system
(5.4.4) to the specified final state $x(N) = x^F$ should be determined.
Optimization is practically omitted in this time interval, since the
state x^F is given. The optimal value S_{N-1} of the performance index J_{N-1},
given the assumed state $x^m(N-1)$, is given by

$$S_{N-1}(x^m(N-1)) = J^m_{N-1}(x^m(N-1)), u^m(N-1)) =$$

$$= L(x^m(N-1), u^m(N-1))+g[x(N)], \quad m=1,...,M^2$$

The corresponding control vector is according to (5.2.24) given by

$$u^m(N-1) = [\hat{B}^T(x^m(N-1))\hat{B}(x^m(N-1))]^{-1}\hat{B}^T(x^m(N-1))[\frac{x(N)-x^m(N-1)}{\Delta t} -$$

$$- \hat{A}(x^m(N-1))]$$

At the (N-j)-th optimization step, the control vectors $u^{m\ell}(N-j)\in U$ should
be determined, which transfer the system from each state $x^m(N-j)\in X$,
m=1,...,M^2 belonging to a set of assumed states, to the states $x^\ell(N-j+1)$,
ℓ=1,...,M^2 which have already been assumed in the previous optimization
step N-j+1. The optimal control vector is the one (between all these
$u^{m\ell}(N-j)\in U$) which minimizes the performance index during the entire
time interval [N-j, N]. The control vector $u^{m\ell}(N-j)$, given the pair of
states $x^m(N-j)$, $x^\ell(N-j+1)$, is given by

$$u^{m\ell}(N-j) = [\hat{B}^T(x^m(N-j))\hat{B}(x^m(N-j))]^{-1}\hat{B}^T(x^m(N-j))[\frac{x^\ell(N-j+1)-x^m(N-j)}{\Delta t} -$$

$$- \hat{A}(x^m(N-j))], \quad m=1,...,M^2, \quad \ell=(m^*-1)M+1,...,m^*M$$

$$m^* = m - [\frac{m-1}{M}]M \qquad (5.4.6)$$

The value of the cost functional is given by

$$J^{m\ell}_{N-j}(x^m(N-j), u^{m\ell}(N-j)) = L(x^m(N-j), u^{m\ell}(N-j)) +$$

$$+ S_{N-j+1}(x^\ell(N-j+1)) \qquad (5.4.7)$$

while the optimal value of the performance index for the transfer from
the state $x^m(N-j)$ onto the optimal second part of the trajectory is

$$S^m_{N-j}(x^m(N-j)) = \min_{u^{m\ell}(N-j)\,\in U} J^{m\ell}_{N-j}(x^m(N-j),\; u^{m\ell}(N-j))$$

At the last optimization step the system state is specified $x(0)=x^o$. For this state we calculate the controls $u(0)\in U$ which transfer the system to all the states $x(1)$ supposed, and choose that control $u^*(0)$ which yields the minimal cost functional on the entire motion execution time $[t_o,\, t_1]$

$$S_o = \min_{u(0)\,\in U} J^*_o(x^o,\; u(0))$$

The optimal trajectory is now reconstructed backward. The optimal control $u^*(0)$ transfers the system from x^o to the optimal state $x^*(1)$. For this state the optimal control $u^*(1)$ has already been determined, together with the optimal state $x^*(2)$, and so on. Thus, we obtain the set of optimal control vectors $u^*(0)$, $u^*(1),\ldots,u^*(N-1)$, which transfer the system along the optimal trajectory x^o, $x^*(1),\ldots,x^*(N-1)$, x^F.

The above method for optimal nominal trajectory synthesis is general in the sense that the performance index can be chosen arbitrarily, as well as the constraints on the state and control vectors.

Let us now consider in more detail the constraints on the state space trajectory $x(N-j)\in X$, $j=0,\ldots,N$, i.e. how the set of vectors $x(N-j)$ assumed should be selected. We will consider only nonredundant manipulators (with n degrees of freedom).

The dynamic programming algorithm allows for the introduction of all the types of constraints encountered in practice. First, there are the constraints imposed on the external coordinate trajectory, resulting in the constraints on the state space trajectory. On the other hand the constraints are frequently imposed directly on the joint coordinates and velocities.

In the domain of the external coordinates $x_e\in X_e\subset R^n$, a region is usually specified, through which the manipulator end-effector should pass between the initial position $x^o_e = f(q^o)$ and the final position $x^F_e=f(q^F)$. These constraints may also take into account the presence of obstacles in the manipulator work space, if such exist. The points in the work space occupied by the obstacles are then excluded from the admissible region of the manipulator hand motion. One can, also, check whether any of the inboard links collides with the obstacle, and if so, then exclude the assumed external coordinate vector from the allowable set of positions. One can also specify a set of external points x_e^k, k=1,...,K through which the manipulator end-effector should pass during its motion between the initial and the final point.

On the other hand, the constraints on joint coordinates usually include the mechanical boundaries on the manipulator motion, $q \in Q \subset R^n$, as well as the constraints on joint velocities, acceleration and jerk (third-order derivatives), thus providing for smooth, jerkless motion.

Now, let us describe how to calculate the state vectors $x^m(N-j)$. In the optimization step (N-j) a set of external coordinate vectors $x_e^i(N-j)$, i=1,...,M should be chosen. The corresponding joint coordinate vectors $q^i(N-j) = f^{-1}(x_e^i(N-j))$ are evaluated for each of them (the inverse transformation will be assumed as unifold, since we are concerned with nonredundant manipulators and the solution in the vicinity of an initial solution is required). For each pair $q^i(N-j)$, i=1,...,M and $q^k(N-j+1)$, k=1,...,M we calculate the joint velocity

$$\dot{q}^m(N-j) = (q^k(N-j+1)-q^i(N-j))/\Delta t,$$

$$i=1,...,M, \quad k=1,...,M, \quad m=(i-1)M+k \qquad (5.4.8)$$

Finally, the state vector is evaluated as

$$x^m(N-j) = [q^{iT}(N-j)\dot{q}^{mT}(N-j)]^T, \qquad m=1,...,M^2 \qquad (5.4.9)$$

The trnasfer from the state $x^m(N-j)$ into the state $x^\ell(N-j+1)$ is feasible only if the state $x^\ell(N-j+1)$ starts exactly from the joint position $q^k(N-j+1)$ which corresponds to the state $x^m(N-j)$. This results in the fact that the set of indices ℓ is constrained to $\ell=(m^*-1)M+1,...$ ...,m^*M, where $m^* = m - [\frac{m-1}{M}]M$.

Let us now consider the dimensionality problem. Generally, for an n degree-of-freedom nonredundant manipulator n variables should be varied (n external coordinates). For a six degree-of-freedom manipulator that would be a severe dimensionality problem, requiring large computer memory and time. For a three degree-of-freedom the optimization can be carried out provided we have a sufficiently large computer memory.

Energy optimal velocity distribution

The optimization problem which frequently occurs in practice is that the manipulator hand path is already specified, while the velocity distribution along the path is not strictly defined. This velocity may become a subject of optimization with respect to some performance index. The dynamic programming technique, described above, may efficiently be applied to this problem, since it requires the variation of a single variable only.

Let us consider an n degree-of-freedom manipulator moving along a specified path in the external coordinate space. The trajectory is specified by its parameter equation

$$x_e = r(\lambda(t)) \qquad (5.4.10)$$

where $x_e \in R^n$ is the external coordinate vector, $r: R \to R^n$ is a given vectorial function, $\lambda \in [0, 1]$ is the scalar parameter, $r(0) = x_e^O = f(q^O)$, $r(1) = x_e^F = f(q^F)$. The time history $\lambda(t)$, $t \in [t_o, t_1]$ is not specified and should be determined in the optimization procedure.

An example of the trajectory (5.4.10) is the straight line trajectory between the initial point x_e^O and the final point x_e^F

$$x_e(t) = x_e^O + \lambda(t)(x_e^F - x_e^O) \qquad (5.4.11)$$

or a circular trajectory (4.5.3), or elliptic trajectory (4.5.4), etc.

In the case when the manipulator end-effector path is given, the set of constraints (5.4.3) becomes

$$x(0) = x^O, \quad x(N) = x^F$$

$$x_e(k) = r(\lambda(k)), \qquad k=1,\ldots,N-1 \qquad (5.4.12)$$

$$|u_i(k)| < u_{imax}, \qquad i=1,\ldots,n, \quad k=0,\ldots,N-1$$

As we have seen, the optimization procedure starts from the final point and goes back to the initial point. Once the optimal trajectory on the interval $[N-j+1, N]$ has been obtained, the optimization is extended to the next interval $[N-j, N]$ in the following way. The set of parameters $\lambda^i(N-j)$, $i=1,\ldots,M$ are selected, which determine the external coordinate vectors $x_e^i(N-j)$, according to (5.4.12), and simultaneously the corresponding joint coordinate vectors $q^i(N-j) = f^{-1}(x_e^i(N-j))$. For each pair of parameters $\lambda^i(N-j)$ and $\lambda^k(N-j+1)$, $i,k=1,\ldots,M$, the corresponding state vector $x^m(N-j)$, $m = (i-1)M + k$ is evaluated according to Equations (5.4.8) and (5.4.9). The control vector $u^{m\ell}(N-j)$ which transfers the system from the state $x^m(N-j)$ to the state $x^\ell(N-j+1)$, $m=1,\ldots,M^2$, $\ell=(m^*-1)M+1,\ldots,m^*M$, is evaluated from Equation (5.4.6). The optimal transfer, starting from the state x^m, is the transfer $m \to \ell$ to which the minimal value of the performance index corresponds. Thus we determine the optimal value of the index ℓ. The whole procedure is repeated for all the values of $j=N,\ldots,1$. For the first and the last time interval the state vectors are already determined by (5.4.12).

The block diagram of the proposed algorithm is shown in Fig. 5.3.

The set of parameters $\lambda^i(k)$, $i=1,\ldots,M$ is selected before the optimization starts. It is convenient to assume an initial distribution $\lambda^o(k)$ (e.g. a linear function between 0 and 1, or a parabolic profile $\hat{\lambda}(t)$), and then form a set of parameters $\lambda^i(k)$ around this initial distribution, by adding and subtracking a fixed increment $\Delta\lambda$

$$\lambda^i(k) = \lambda^o(k) + (i - [\tfrac{M+1}{2}])\Delta\lambda, \qquad i=1,\ldots,M$$

The optimization procedure yields the optimal distribution $\lambda^*(k)$. These optimal values of $\lambda^*(k)$ must not reach the boundaries of the set $\lambda^i(k)$, $k=0,\ldots,N$, i.e. the following inequality must hold

$$\lambda^1(k) < \lambda^*(k) < \lambda^M(k), \qquad k=0,\ldots,N$$

If this is not the case the number M of the assumed parameters at each optimization interval should be increased, or alternatively, the initial distribution $\lambda^o(k)$ should be modified. Practical experiments showed that $M \leqslant 10$ is quite sufficient to obtain the optimal velocity profile.

Let us now consider a particular case when the robot joints are powered by DC motors, modelled by the second order linear models (5.2.4), (5.2.8) - (5.2.9). If the energy consumption of the actuators is considered as the cost functional, the performance index (5.4.2) becomes

$$J = \sum_{i=1}^n \int_{t_o}^{t_1} u_i i_i \, dt = \int_{t_o}^{t_1} (u^T Q_1 u + x^T Q_2 u) \, dt \qquad (5.4.13)$$

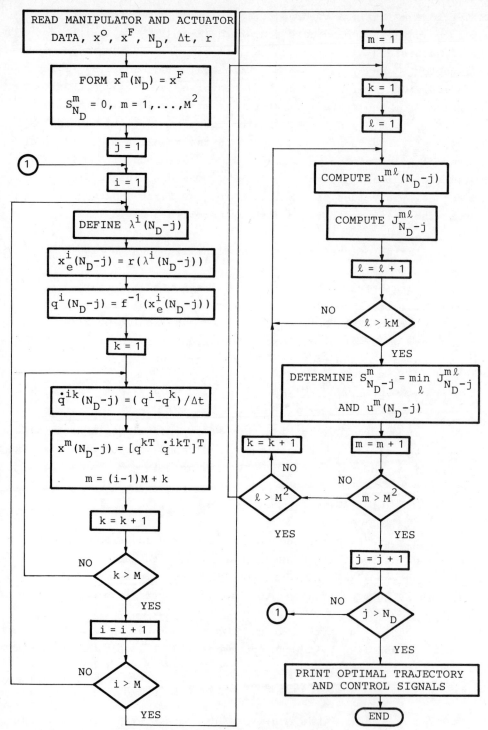

Fig. 5.3. Flow-chart of the algorithm for determining optimal velocity distribution

where u_i is the input voltage of the ith motor, i_i - the rotor current, $Q_1 \in R^{n \times n}$, $Q_2 \in R^{2n \times n}$. The matrix Q_1 is the diagonal matrix $Q_1 = \text{diag}(1/R_1, \ldots \ldots, 1/R_n)$, where R_i, $i=1,\ldots,n$ are the rotor resistances. All the elements of the matrix Q_2 are equal to zero, except for the elements $q_{2(n+i),i} = -K_{mei} N_{vi}/R_i$, $i=1,\ldots,n$, where K_{mei} is the mechanical-electrical constant of the motor i, N_{vi} is the angular velocity reduction ratio.

The functions $L(x, u)$, $g[x(t_1)]$, corresponding to the cost functional (5.4.13) in the discretized form, are given by

$$L(x(k), u(k)) = \Delta t[u^T(k)Q_1 u(k) + x^T(k)Q_2 u(k)]$$

(5.4.14)

$$g[x(N)] = 0$$

The above described method for energy optimal motion synthesis is applicable to the evaluation of the energy optimal velocity distributions for electrically driven, nonredundant manipulators.

An example of the energy optimal velocity profile synthesis

In order to illustrate the algorithm and to analyze the results obtained, we will consider the UMS-1 manipulator in its basic configuration, i.e. with three degrees of freedom (Fig. 5.4). The manipulator dynamic

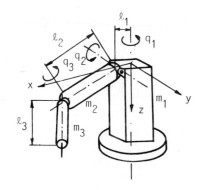

LINK	m_i [kg]	ℓ_i [m]	J_N [kgm^2] [*]	J_S [kgm^2] [*]
1	15.5	0.14	–	0.16
2	2.45	0.35	0.035	0.0038
3	0.7	0.36	0.0048	0.0003

[*] J_S and J_N denote the moments of inertia about longitudinal and transversal axes of the links

Fig. 5.4. The UMS-1 manipulator with 3 degrees of freedom

parameters are shown in Fig. 5.4, while the elements of the actuators' model matrices A^i, b^i and f^i are shown in Table T.5.1. The robot joints

are powered by DC motors modelled by the second-order linear model
(5.2.4), (5.2.8)-(5.2.9). The manipulator end-effector is taken into
account as an effective load at the end of the third link.

i	a_{22}^i	b_2^i	f_2^i
1,2	-10.8	1.42	-0.66
3	-15.2	10.7	-30.

Table T.5.1. The UMS-1 actuators' parameters

The energy-optimal velocity distribution for the straight line manipulator movement between the points $x_e^o = [x^o \; y^o \; z^o]^T = [0.43 \; 0.4 \; 0.18]^T$
[m] and $x_e^F = [0.33 \; 0.48 \; 0.25]^T$ [m], in the time $t_1-t_o = 0.5s$, is considered. The optimization interval is taken to be 20 ms. In this example it is assumed that the manipulator starts from rest and should be
stopped at the final point.

The obtained optimal velocity distribution $\dot{\lambda}(t)$ is depicted in Fig. 5.5.

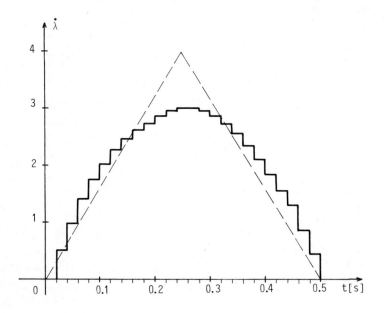

Fig. 5.5. The optimal velocity distribution $\dot{\lambda}(t)$

As we can see, it is very similar to a parabolic velocity profile. The
corresponding optimal external coordinate velocities, being propor-

tional to $\dot{\lambda}(t)$

$$\dot{x}_e(t) = \dot{\lambda}(t) (x_e^F - x_e^O)$$

are shown in Fig. 5.6. For the sake of comparison the triangular velocity profiles are also shown in these figures. The corresponding time

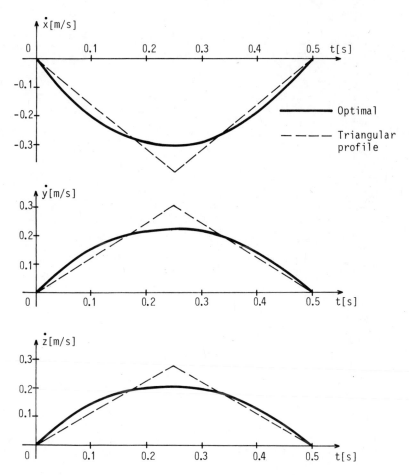

Fig. 5.6. The Cartesian velocities for the optimal and the
triangular velocity profile

histories of the joint rates are illustrated in Fig. 5.7. The nominal
driving torques and motor control signals are shown in Figures 5.8 and
5.9, respectively. The comparison of the energy consumption of the actuators for the optimal motion and the triangular velocity profile, is
illustrated in Fig. 5.10.

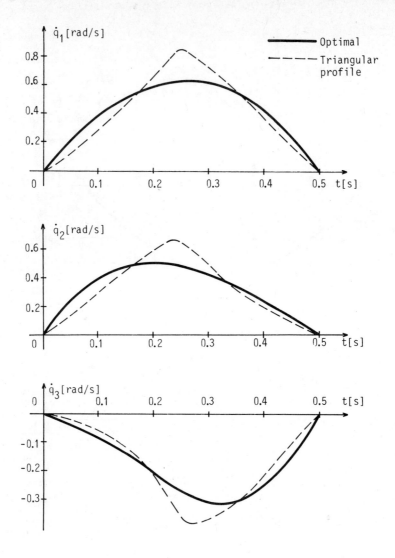

Fig. 5.7. The joint rates for the optimal and triangular
velocity profiles

The results obtained in the process of energy optimization of the velocity distribution, using the complete dynamic model of the system,
point to the following conclusions.

The energy favourable manipulator motion is obtained when the speed has
a steep slope at the beginning, and then a slower increase up to a maximal magnitude. On the second part of the trajectory the deceleration
is slower first, and then increases. The velocity profile, which is
most similar to the optimal profile, is a parabolic one. It is evident

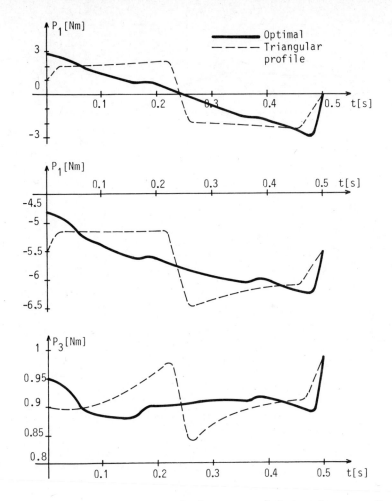

Fig. 5.8. The time-history of the motor driving torques for the
optimal and triangular velocity profiles

from Fig. 5.5 that a better approximation of the optimal distribution
$\dot{\lambda}(t)$ would be obtained either by increasing the number of assumed values for λ at each optimization step, or by decreasing the time interval
Δt. However, further improvement of the optimal solution would not result in considerable energy savings, but only in further smoothing of
the curves from Figures 5.8 and 5.9.

Figure 5.10 indicates that the energy saving in this example amounts to
18%. Such energy savings may represent considerable savings in the case
when manipulation of heavy objects is performed. Energy-optimal velocity generation, not only saves energy, but also minimizes the strains of
the actuators and the mechanism itself, since abrupt variations of the

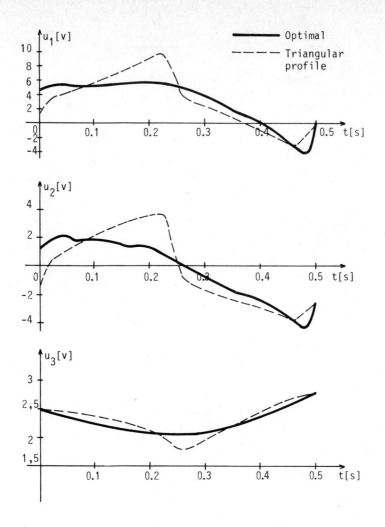

Fig. 5.9. The time-history of the control signals for the optimal and the triangular velocity profiles

acceleration are also unsuitable from the standpoint of energy consumption.

The proportional saving with respect to some suboptimal velocity profile (e.g. triangular or trapezoidal) depends a great deal on the motion execution time. Figure 5.11 illustrates this statement. It depicts the total energy consumption versus motion execution time, for the UMS-1 manipulator motion between points $x_e^O = [0.49\ 0.36\ 0]^T$ [m] and $x_e^F = [0.44\ 0.47\ 0.1]^T$ [m], with the optimal velocity (the solid line) and the triangular velocity (the dashed line). For both profiles there exists a range of the execution times where the energy consumption is practi-

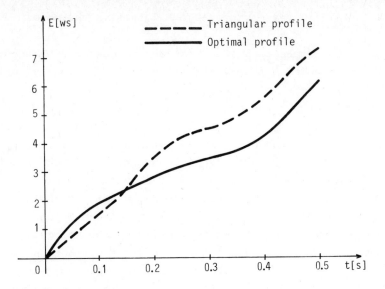

Fig. 5.10. The time-history of the energy consumption for the
optimal and the triangular velocity profiles

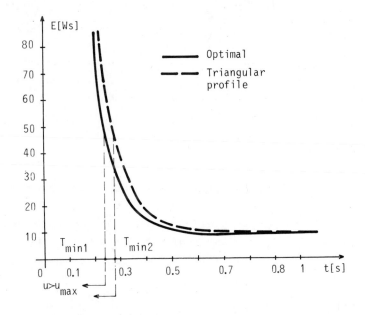

Fig. 5.11. Energy consumption versus motion execution time,
for the optimal and the triangular velocity profiles

cally constant. These motion execution times may be considered as energy optimal. With decreasing the motion execution time, the total energy consumption rapidly increases. For long execution times the energy

consumption slowly increases, due to the losses in the motor.

The proportional difference between the energy consumed by the optimal
and triangular velocity, is very small for longer motion execution
times. With the decrease of the time the difference increases up to 30%,
where the constraints on the motor input voltages are violated. This
violation occurs for longer execution times in the case of the triangular velocity than in the case of the optimal one ($T_{min1} > T_{min2}$ in
Fig. 5.11). This is quite understandable, when we bear in mind that the
dynamic programming algorithm seeks for the optimal solution between
all the possible motions, which simultaneously satisfy the constraints
on the actuators' capabilities, given the motion execution time. Therefore, by applying the algorithm for obtaining the energy optimal velocity distribution, given a trajectory, for various motion execution
times, one can obtain the minimal time in which the movement is still
achievable. This time, however, is not an absolute time optimum, since
the cost functional was the energy consumption. It was evaluated suboptimally together with the energy optimal condition.

As we have already mentioned, the method for optimal velocity generation, described in this section, is off-line in character, since it is
based on the dynamic programming. However, the conclusions drawn from
the optimal synthesis, may also be utilized in on-line motion generation. Namely, the parabolic velocity distribution is easily achievable
in real-time, and at the same time represents a very good approximation
of the energy optimal velocity profile (the difference in the energy
consumed is about 1%).

As we have observed, the optimization method does not depend on the dynamic modelling technique employed, i.e. any of the methods based on
Newton-Euler, Lagrange, Gibss-Appel equations, may be used. However, it
is convenient to employ a computationally efficient method in order to
reduce the optimization time (the most convenient are the analytical
dynamic models [16-18]). In the case when the analytical model is used
in the optimal velocity synthesis, the optimization lasts several mi-
nutes (on the minicomputer PDP 11/70, FORTRAN IV plus), which is an ac-
ceptable time for the off-line trajectory optimization.

5.5. Quasioptimal Nominal Trajectory Synthesis Using Decentralized System Model

The methods for the optimal nominal trajectory synthesis, described in the previous two sections, are intended for off-line motion generation, either because of the long optimization time or because of the nature of the optimization algorithm itself. On the other hand, however, the significance of the real-time motion generation increases every day, especially for the manipulation in variable or unknown environments. In this section we will present a method for quasioptimal nominal trajectory synthesis, which is implementable in real time.

As we saw in Section 5.3, the optimal nominal trajectory synthesis based on the complete dynamic model of the system (5.3.1), by solving the canonic system of equations (5.3.3)-(5.3.6), is burdened by substantial practical difficulties, such as problems with equations' integration, high dimensionality of the system, evaluation of the partial derivative matrices, etc). On the other hand, solving the optimization problems according to the dynamic programming or linear programming algorithms suffers from dimension problems, resulting in the relatively long computational time. In this section a method for suboptimal nominal trajectory synthesis will be proposed, employing a simplified dynamic model of the system, and which makes the simple application of the classical optimal control theory possible. The closed-form solution obtained is suitable for the real-time optimization of the manipulator motion [75, 79-80].

The decentralized, simplified dynamic model of an n degree-of-freedom open active mechanism will be formed in the following way. Each of the degrees of freedom is considered as an independent subsystem modelled by the second-order linear model

$$\dot{x}^i = A^i x^i + b^i u_i + f^i P_i, \quad x^i(0) = x^{io}, \quad x^i(T) = x^{iF},$$

$$i=1,\ldots,n \qquad (5.5.1)$$

where $x^i = [x^i_1 \ x^i_2]^T = [q_i \ \dot{q}_i]^T \in R^2$ is the state vector of the i-th subsystem, $i=1,\ldots,n$, $A^i \in R^{2 \times 2}$, $b^i \in R^2$, $f^i \in R^2$ are the system model matrices, P_i - the driving torque (force) acting on the output reducer shaft at joint i, $u^i \in R$ the control variable of the i-th actuator. The torque will be approximated by

$$P_i = H^e_{ii} \ddot{q}_i + h^e_i, \qquad i=1,\ldots,n \tag{5.5.2}$$

where H^e_{ii} is an equivalent moment of inertia of the i-th degree of freedom for the given manipulator trajectory, while h^e_i represents an equivalent torque due to the force of gravity. The parameters H^e_{ii} and h^e_i are adopted to be constant for the given movement between the two end-points. Equation (5.5.2) is an approximation of the real torque in joint i, since it takes into account the acceleration of the link i only (while the influence of the other links is neglected), and for the reason that H^e_{ii} and h^e_i are constant parameters (independent of the joint coordinates and rates along the trajectory). The equivalent moment of inertia H^e_{ii} may be predetermined, for example, in the following way: the influence of the masses of links $i+1,\ldots,n$ at average distances from the i-th link, are added to the eigen moment of inertia of link i, given the initial and final trajectory point. The torque h^e_i may be determined in a similar may.

By substituting the expression (5.5.2) into the actuator model (5.5.1), the model of the subsystem i is obtained in the form

$$\dot{x}^i = A^i_1 x^i + b^i_1 u_i + f^i_1, \qquad i=1,\ldots,n \tag{5.5.3}$$

where $A^i_1 \in R^{2 \times 2}$, $A^i_1 = (I_2 - f^i_H)^{-1} A^i$, $f^i_H = [0 \mid f^i H^e_{ii}]$; $b^i_1 \in R^2$, $b^i_1 = (I_2 - f^i_H)^{-1} b^i$; $f^i_1 \in R^2$, $f^i_1 = (I_2 - f^i_H)^{-1} f^i h^e_i$. The system (5.5.3) is the second-order, linear, stationary system.

Let us consider the energy performance index (5.4.13) for the subsystem i

$$J(u_i, x^{io}) = \int_0^T (u^T_i Q^i_1 u_i + x^{iT} Q^i_2 u_i)\,dt \tag{5.5.4}$$

where $Q^i_1 \in R > 0$, $Q^i_2 \in R^2$ and T is a fixed motion exectution time.

In terms of the linear, decentralized models (5.5.3) the optimization problem can be summarized thus: it is necessary to determine such control signals $u_i(t)$, $t \in [0, T]$, $i=1,\ldots,n$ which transfer the subsystems (5.5.3) from the initial states $x^i(0) = x^{io}$ to the final states $x^i(T) = x^{iF}$, and simultaneously minimize the cost functional (5.5.4).

The solution to the above optimization task is obtained in the case when the manipulator joints are powered by DC motors, modelled by the second-order models (5.2.4), where the model matrices

$$A^i = \begin{bmatrix} 0 & 1 \\ 0 & \alpha_i \end{bmatrix}, \quad b^i = \begin{bmatrix} 0 \\ \beta_i \end{bmatrix}, \quad f^i = \begin{bmatrix} 0 \\ \gamma_i \end{bmatrix} \tag{5.5.5}$$

are given by (5.2.8)-(5.2.9). The matrices of the energy performance index (5.5.4) are

$$Q_1^i = 1/R_i, \quad Q_2^i = [0 \; g^i]^T = [0 \; -K_{me}^i N_{vi}/R_i]^T \tag{5.5.6}$$

The canonical system of equations (5.3.3)-(5.3.5) for the system (5.5.3) is given by

$$\dot{x}^i = A_1^i x^i + b_1^i u_i + f_1^i, \quad x^i(0) = x^{io}, \quad x^i(T) = x^{iF}$$

$$\dot{p}^i = -Q_2^i u_i - A_1^i p^i \tag{5.5.7}$$

$$2Q_1^i u_i + x^{iT} Q_2^i + p^{iT} b_1^i = 0$$

where $p^i = [p_1^i \; p_2^i]^T$ is the adjoint vector for the subsystem i. By substituting the matrices A_1^i, b_1^i, f_1^i and (5.5.6) into the Equations (5.5.7) the problem is reduced to a set of scalar differential equations. The optimal solution is obtained in the closed form

$$q(t) = q_i^o + c_1^i \frac{e^{\lambda_1 t} - 1}{\lambda_i} + c_2^i \frac{1 - e^{-\lambda_i t}}{\lambda_i} + c_3^i t \tag{5.5.8}$$

where the parameter λ_i is determined from

$$\lambda_i = \sqrt{r^i}, \quad r^i = a_1^{i2} + a_2^i a_4^i$$

$$a_1^i = c^i - d^i g^i R_i/2, \quad a_2^i = -d^{i2} R_i/2, \quad a_4^i = g^{i2} R_i/2 \tag{5.5.9}$$

$$c^i = \alpha_i/(1 - \gamma_i H_{ii}^e), \quad d^i = \beta_i/(1 - \gamma_i H_{ii}^e)$$

The constants c_1^i, c_2^i and c_3^i are obtained from the boundary conditions in the initial and the final points. They satisfy the linear system

$$c_1^i + c_2^i + c_3^i = \dot{q}_i^o$$

$$c_1^i e^{\lambda_i T} + c_2^i e^{-\lambda_i T} + c_3^i = \dot{q}_i^F \tag{5.5.10}$$

$$c_1^i \, \frac{e^{\lambda_i T} - 1}{\lambda_i} + c_2^i \, \frac{1 - e^{-\lambda_i T}}{\lambda_i} + c_3^i T = q_i^F - q_i^O$$

The optimal control signals which correspond to the optimal trajectories (5.5.8) can also be obtained. However, they would not lead the real system along the optimal trajectory since the simplified model was considered. The control should be determined from the complete model, or the joint coordinates obtained may directly be used as input signals for positional servosystems.

From the standpoint of the complete system the obtained trajectories are suboptimal, since the decentralized, simplified system model was employed.

Example of the quasioptimal nominal trajectory

The above described procedure for quasioptimal trajectory synthesis will be illustrated by an example. The motion of the UMS-1 manipulator will be optimized (Fig. 5.1). The actuator parameters are shown in Table T.5.1.

Motion between the initial point $x_e^O = [0.49 \; 0.36 \; 0]^T$ [m] and the final point $x_e^F = [0.44 \; 0.47 \; 0.1]^T$ [m] in the time $T = 0.5$ s is considered. The initial and final velocity of the manipulator tip is taken to be zero. The joint coordinates $q^O = f^{-1}(x_e^O)$ and $q^F = f^{-1}(x_e^F)$ are evaluated first. Then, the equivalent moments of inertia are estimated $H_{11}^e = 0.66$, $H_{22}^e = 0.2$, $H_{33}^e = 0.043$ [kgm^2]. The moment of inertia H_{11}^e is, for example, evaluated in the following way. The impact of the masses of the second and the third link, being at the average distances from the first link, are added to the eigen moment of inertia of the first link. Then, the parameters λ^i and c_1^i, c_2^i, c_3^i, $i=1,2,3$ are evaluated according to (5.5.9) - (5.5.10).

The quasioptimal joint trajectories obtained, $q_i(t)$, $i=1,2,3$ together with the corresponding Cartesian coordinates of the manipulator tip, $x(t)$, $y(t)$, $z(t)$, are shown in Fig. 5.12. For the sake of comparison, the straight-line Cartesian trajectory and the corresponding joint trajectory are shown by the dotted lines in the same figure. Motion along the straight line is carried out with the velocity which is optimal with respect to the total energy consumption of the actuators

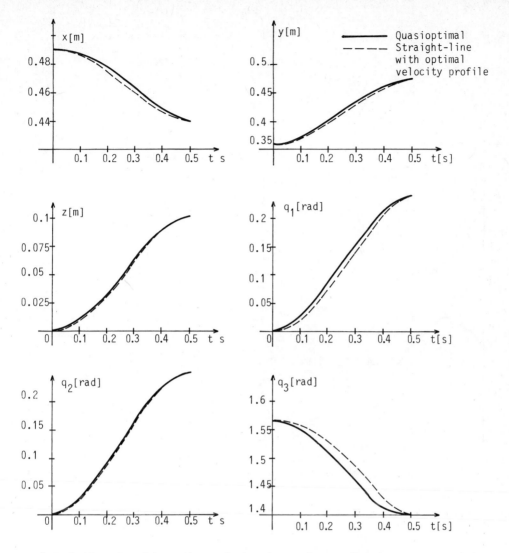

Fig. 5.12. Time histories of the Cartesian and joint coordinates
for the quasioptimal motion and the straight-line
motion with the energy optimal velocity

(obtained by the previously described dynamic programming algorithm).
The joint rates and the corresponding control signals are shown in Fig.
5.13. The control signals transfer the system along the trajectories.
It is determined on the basis of the complete dynamic model of the system. Figure 5.14 depicts the time history of the total energy consumption for the quasioptimal and the straight-line motion with the optimal
velocity.

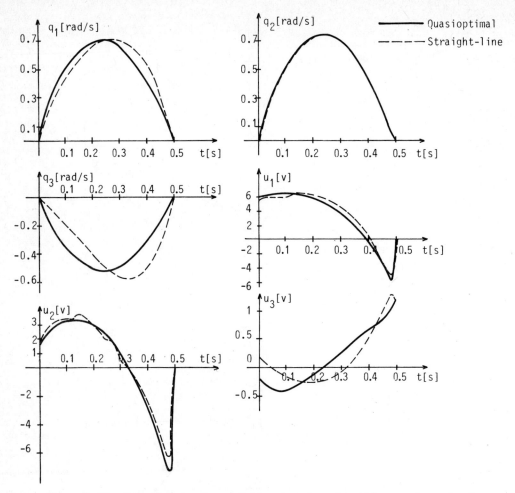

Fig. 5.13. Time histories of the joint velocities and control
signals for the quasioptimal and the straight line
trajectory with the optimal velocity

The comparison between the quasioptimal trajectory and the straight-
-line motion with the energy optimal velocity, indicates that these results are in goad accordance. The slightly greater difference apppear-
ing with the third joint rate is the consequence of the fact that the
quasioptimal trajectory does not provide for straight line motion. The
comparison of the energy consumed indicates that only very small differences exist between the two total energy consumptions.

Bearing in mind the relative simplicity of the above quasioptimal motion
synthesis, together with good optimization results, one can conclude
that it can efficiently be used for on-line, as well as for off-line

motion synthesis.

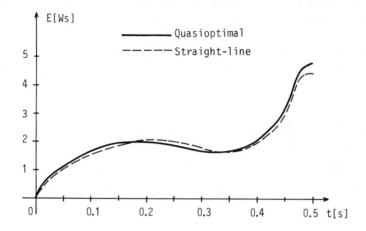

Fig. 5.14. The energy consumed along the quasioptimal and the straight-line trajectory with the optimal velocity profile

5.6. Summary

In this chapter we have dealt with the manipulator motion synthesis, where the system is considered as a dynamic system modelled by the complete, nonlinear dynamic model of the mechanism and the actuators. The complexity of the solution to the general optimal control problem for such systems, was illustrated. Regarding the practical importance of the energy optimal motion synthesis which simultaneously provides for smooth, jerkless motion and minimal actuators' strains, a particular attention was paid to the energy optimal motion of nonredundant manipulators. An algorithm for determining the energy optimal velocity distribution, given a manipulator end-effector path, was presented. The optimal profile obtained is close to the parabolic velocity distribution. Another algorithm presented refers to the energy quasioptimal trajectory synthesis, given the two end points. It is computationally simple, being implementable in real time.

Chapter 6
Motion Generation for Redundant Manipulators

6.1. Introduction

The requirements involving manipulator robot flexibility and adaptivity
are increasing every day. The use of the greater number of degrees of
freedom, with respect to the minimal number required for the completion
of a certain class of task, plays an important role in improving the
robot flexibility. This is especially important in manipulation in
cluttered work space, where the increased number of degrees of freedom
may efficiently be used for avoiding obstacles, or avoiding the manipulator singularities. Therefore, research becomes more and more significant in the domain of redundant manipulator control, aimed at utilizing the redundant manipulator preferences completely.

In this chapter we shall consider the problem of nominal trajectory
synthesis for redundant manipulators. At this control level the redundancy problem is completely resolved and the nominal joint trajectories
are obtained. Once they have been determined, the problem of the tracking
of these trajectories in the presence of various disturbances is reduced to the same problem as that of the nonredundant manipulators, and
thus the application of all the control schemes elaborated for nonredundant manipulator control is made possible.

If one is dealing with redundant manipulator motion synthesis in an
environment free of obstacles, the end-effector trajectory may be specified in the same way as for nonredundant manipulators, i.e. in such a
way as to provide for the functionality of motion. However, with redundant manipulators, once the end-effector trajectory is specified, the
joint motion is not uniquely determined, since there is an indefinite
number of solutions to the inverse kinematic problem. The presence of
a surplus of degrees of freedom can efficiently be utilized for achieving
an optimal motion with respect to some performance index, together with
achieving the desired gripper positions and orientations. In the next
section we will present an overview of the methods, purely kinematic in
nature, which are aimed at resolving the inverse problem ambiguity, i.e.

which are based on the application of kinematic performance criteria.

In Section 6.3 we will present a method for energy optimal motion synthesis for redundant manipulators, based on the inverse problem solution being optimal with respect to the energy consumption of the actuators. This performance index takes into account the dynamic model of the system.

The motion synthesis problem in an environment with obstacles, represents a separate, but very significant problem. The redundant manipulators play an important role in its solution. The problem of the automatic planning of manipulator transfer movements in environments with obstacles includes several subproblems, such as obstacle modelling, handling sensors' information concerning obstacles, searching for collision-free paths, etc.

Generally, a control algorithm should provide that during the motion between the initial and the final point, none of the links nor the payload collides with obstacles, while the deviation from the desired trajectory is minimized. This is not easily achievable, especially in obstacle-cluttered workspace. Considering the manipulator links' movements simultaneously with the manipulator hand trajectory modification, increases the complexity of the solution. This is the reason why, most frequently, two separate problems have been considered in literature:

1. given an initial and final location and orientation of a work object, and a set of obstacles located in space, the problem is to find a continuous path for the object, from the initial position to the goal position, so that collisions between obstacles and the object along the path are avoided;

2. given an initial manipulator configuration, desired manipulator end--effector trajectory (positions and orientations), and a set of obstacles located in space, the problem is to determine such a joint coordinates' evolution which enables the manipulator hand to move along the desired trajectory, while the manipulator links stay away from the obstacles and the joint coordinates stay within their constraints.

The first problem - that of finding the path, is very complex, and has been solved in several ways [53-56]. This problem is not directly related to the redundant manipulators, so that we shall not consider it in this book.

In contrast, the second problem is inherently associated with the redundant manipulators, since the surplus of the degrees of freedom may efficiently be utilized for avoiding the obstacles by the inboard links, if the end-effector path is given. An overview of the methods referring to this subject will be presented in Section 6.4. An algorithm for determining such a motion for the manipulator joints, providing for obstacle avoidance, and achieving the desired manipulator hand positions and orientations, will be described in Section 6.5. The motion is determined optimally with respect to a performance index, taking the distance between the obstacle and the point on the arm which is closest to the obstacle into account. The algorithm is computationally efficient, and real-time implementable.

6.2. Kinematic Methods for Redundant Manipulator Motion Generation

As we have mentioned in the introduction to this chapter, the central problem in redundant manipulator motion generation is the inverse kinematic problem. We will assume in this discussion, that the end-effector trajectory is defined by the manipulation task, i.e. by the external coordinate trajectory $x_e(t)$, $t \in [0, T]$. The external coordinate vector $x_e \in X_e \subset R^m$ is an m-dimensional vector describing the manipulator hand position and orientation (completely or partially). One should determine the differential change of joint coordinates $\Delta q \in R^n$, given the differential change of external coordinates $\Delta x_e \in R^m$, from the linear system of m equations in n unknowns

$$J(q) \Delta q = \Delta x_e \qquad (6.2.1)$$

where $J(q): R^n \to R^{m \times n}$, $m < n$. Since we are dealing with redundant manipulators, the number of external coordinates describing the task, is less than the number of degrees of freedom. Here, the question of the existence of the solution arises.

When the Jacobian matrix $J(q)$ is a full rank matrix, i.e. rank $J = m$, there is an indefinite number of solutions to the system (6.2.1). In order to obtain a unique solution, an additional cost functional should be introduced. This subject will be discussed in the text to follow. However, in the singular points, where rank $J < m$, the existence of the

solution depends on the vector Δx_e (Cronecker-Capelli theorem). If
$\text{rank} J = \text{rank}[J \mid \Delta x_e] < m$, there still exists an unlimitted number of solutions. If rank $J \neq \text{rank}[J \mid \Delta x_e]$ holds, there is no solution to the
system, i.e. the given movement can not be achieved.

Most of the methods for overcomming the ambiguity of the inverse problem solution are kinematic in nature, i.e. use the kinematical performance indices [25, 33, 81–88]. We will here present the main results attained in this field so far, which refer to the redundant manipulator motion
synthesis in the absence of obstacles.

There are several ways of removing the indefiniteness of the inverse
problem solution. The simplest way is to reduce computational complexity by commanding only a limited number of joints. In that case n-m elements of the vector Δq are fixed to zero (assuming that rank J=m). If
the set of variables to be changed was selected simply, it would easily
happen that the potential feasibilities of redundant manipulators were
not utilized. The selection of the variables q_i which will be varied
(from the C_n^m possible solutions), may be carried out with respect to
the criterion of the maximum determinant of the corresponding m×m minor
of J [81–82]. However, this would require relatively long computation
time. When a joint reaches its mechanical lock it should be excluded
from the set of varied coordinates.

Most frequently, however, the ambiguity is overcome by minimizing the
performance index $\Omega = \Omega(\Delta q)$, or equivalently $\Omega = \Omega(\dot{q})$ [25, 33, 83]. Various indices have been proposed so far [33], but the most appropriate
and the most frequently used criterion is the quadratic criterion

$$\Omega(\Delta q) = \frac{1}{2}\Delta q^T M \Delta q \qquad (6.2.2)$$

where $M \in R^{n \times n}$ is a symmetric positive definite matrix. This performance
index corresponds to the weighted norm of the vector Δq. It is convenient since it yields the individual solution for the differential increment Δq, which satisfies the linear system (6.2.1). This solution
can be obtained if the Lagrangian of the system is formed

$$L(\Delta q, p) = \frac{1}{2}\Delta q^T M \Delta q + p^T(\Delta x_e - J \Delta q) \qquad (6.2.3)$$

where $p \in R^m$ is the Lagrange multiplier vector. By imposing the conditions that the partial derivatives of the Lagrangian with respect to

Δq and p have to be equal zero at the optimal point Δq^*, p^*

$$\frac{\partial L(\Delta q, p)}{\partial \Delta q}\bigg|_{\Delta q^*, p^*} = 0, \qquad \frac{\partial L(\Delta q, p)}{\partial p}\bigg|_{\Delta q^*, p^*} = 0$$

one obtains the optimal solution in the form

$$\Delta q = G \Delta x_e \qquad\qquad (6.2.4)$$

where $G \in R^{n \times m}$ is the generalized inverse of the Jacobian matrix, given by

$$G = M^{-1} J^T (J M^{-1} J^T)^{-1} \qquad\qquad (6.2.5)$$

The generalized inverse G satisfies the condition $JGJ = J$ [90-92].

The weighting matrix M may be chosen in different ways [25, 33, 83]. In [25] matrix M was chosen so as to emphasize the role of some components of Δx_e, i.e. the matrix M(q) was adopted in the form

$$M(q) = J^T(q) B J(q) \qquad\qquad (6.2.6)$$

This means that, in fact, the optimality criterion

$$\Omega = \frac{1}{2} \Delta x_e^T B \Delta x_e$$

was considered instead of (6.2.2). Here $B \in R^{m \times m}$ is a positive definite matrix. In this case, however, the matrix M(q) is not necessarily positive definite. In [33] M is assumed to be a diagonal matrix, with the elements being dependent on joint coordinates. Most frequently, however, the matrix M is adopted to be a unit matrix $M = I_n$ [27, 83, 85], yielding the minimum norm solution

$$\Delta q = J^T (J J^T)^{-1} \Delta x_e \qquad\qquad (6.2.7)$$

Having the simplest form, this solution is convenient for real time applications. It should be provided, however, that the Jacobian matrix is a full rank matrix.

Beside the quadratic optimality criterion (6.2.2), the use of linear criteria has also been proposed [33]. For example, the criterion of the minimum increment of the joint coordinates was proposed

$$\Omega(\Delta q) = \sum_{i=1}^{n} a_i |\Delta q_i| \tag{6.2.8}$$

as well as the time criterion

$$\Omega(\Delta q) = \max_{i=1,\ldots,n} (|\Delta q_i|/v_i) \tag{6.2.9}$$

where v_i is the given joint rate for the ith coordinate. However, these criteria do not yield unique solutions for Δq (but finite sets of solutions) [33]. For this reason, they are not convenient for nominal trajectory synthesis for redundant manipulators. The method of undetermined coefficients [82, 86] is a similar case, where the solution is obtained as a function of C_n^{m+1} arbitrary coefficients.

If we give up the minimum norm solution, it is possible to introduce a global optimality criterion $H(q)$, beside the local criterion $\Omega(\Delta q)$ [81, 88-89]. Then, an orthogonal component is added to the minimum norm solution

$$\Delta q = G_1 \Delta x_e + (G_2 J - I_n) y \tag{6.2.10}$$

where G_1 and G_2 are the generalized inverse matrices of J, satisfying $JG_1J = J$, $JG_2J = J$, $y \in R^n$ is an arbitrary vector, and I_n is the nth order identity matrix. The vector $(G_2J-I_n)y$ represents the projection of y on the null-space of J [90-91]. If the vector y is adopted as

$$y = \alpha \nabla H(q) \tag{6.2.11}$$

where α is a real scalar, $\nabla H(q)$ is the gradient of the smooth function $H(q)$ which should be minimized. If the projection operator (G_2J-I_n) has been well chosen, then the component $\alpha(G_2J-I_n)\nabla H(q)$ forces $H(q)$ to decrease.

The function $H(q)$ can be selected to match certain motion synthesis objectives (e.g. to avoid external obstacles, to stay away from the mechanical locks of the joints). In [81] an example of motion synthesis, with $H(q)$ being

$$H(q) = \frac{1}{n} \sum_{i=1}^{n} (\frac{q_i - a_i}{a_i - q_{iM}})^2, \qquad a_i = \frac{q_{im} + q_{iM}}{2} \tag{6.2.12}$$

has been considered, where q_{im} and q_{iM} designate the minimum and maximum values of the ith joint coordinate, respectively. In [88] the higher order norms have been proposed.

The above general solution (6.2.10) provides for the minimization of an arbitrary performance index. However, it may at time be inefficient, when the local and global criteria collide, so that $H(q)$ no longer decreases. In that case, a higher control level should exclude the global criterion from the solution and leave only the local criterion.

In [89] the same method for optimizing the redundant manipulator motion was used in order to avoid manipulator singularities, i.e. to keep the manipulator dexterity as high as possible. The global criterion $H(q)$ is adopted as

$$H(q) = \sqrt{\det(J\,J^T)} \qquad (6.2.13)$$

In this case, however, the evaluation of the gradient $\nabla H(q)$ is rather complex. If the following notation is introduced

$$R = J\,J^T = [r_{ij}], \qquad i,j=1,\ldots,m$$

$$y = \alpha\nabla H(q) = [y_1 \cdots y_n]^T$$

then the components $y_1,\ldots,y_n$ are evaluated from

$$y_\ell = \frac{1}{2\sqrt{\det R}} \sum_{i,j=1}^{m} \Delta_{ij}\;_\ell r'_{ij} =$$

$$= \frac{1}{2}\sqrt{\det R}\sum_{i,j=1}^{m} [R^{-1}]_{ij}(_\ell J'_i\,J^T_j + _\ell J'_j\,J^T_i)$$

where

Δ_{ij} - is the cofactor of R,

$$_\ell r'_{ij} = \frac{\partial(r_{ij}(q))}{\partial q_\ell} \quad ,$$

$[R^{-1}]_{ij}$ - is the i,j element of the inverse of R

J_i - the ith row of J

$_\ell J'_i$ - the derivative of J_i with respect to q_ℓ.

As we see, this motion synthesis also requires the evaluation of the partial derivatives of all the Jacobian matrix elements with respect to the joint coordinates.

The general solution to the underdetermined system (6.2.1) can also be obtained by the Moore-Penrose inverse of matrix J (the pseudoinverse J^+) [93-95]. This approach to solving the inverse kinematic problem for redundant manipulators was considered in [87-88]. Its advantage over the generalized inverse approach, is that it yields the inverse even when rank J<m. However, in the vicinity of some singularities the application of the pseudoinverse may also yield unacceptably high joint rates [103]. Besides, from real-time motion generation standpoint, it is inconvenient, owing to relatively high computational complexity [36]. In [63] the singular value decomposition theorem was used in order to derive a canonical, partially decoupled form of kinematic equations. We will here briefly outline this method.

According to the singular value decomposition theorem [93-95], a real m×n matrix J can be represented by

$$J = PSQ^T \qquad\qquad (6.2.14)$$

where $P \in R^{m \times m}$ and $Q \in R^{n \times n}$ are orthogonal matrices and

$$S = \left[\begin{array}{c|c} O_{m-r,n-r} & O_{m-r,r} \\ \hline O_{r,n-r} & S_r \end{array}\right]$$

$$S_r = \text{diag}(s_1, \ldots, s_r), \qquad\qquad s_i > 0, \quad i = 1, \ldots, r$$

$$r = \text{rank } J$$

$O_{i,j}$ - denotes the i×j null matrix.

The pseudoinverse matrix J^+ is now determined as

$$J^+ = QS^+P^T \qquad\qquad (6.2.15)$$

where

$$S^+ = \left[\begin{array}{c|c} O_{n-r,m-r} & O_{n-r,r} \\ \hline O_{r,m-r} & S_r^{-1} \end{array}\right]$$

$$S_r^{-1} = \text{diag}(1/s_1, \ldots, 1/s_r).$$

Using this technique, the kinematic equations (6.2.10) can be presented in a partially decoupled, canonical form. By substituting (6.2.14) and (6.2.15) into (6.2.10), and by premultiplying by Q^T (bearing in mind the orthogonality of matrices P and Q), one obtaines

$$Q^T \Delta q = S^+ P^T \Delta x_e + (I_n - S^+ S) Q^T y \qquad (6.2.16)$$

Upon introducing the notation

$$\Delta v = Q^T \Delta q$$

$$\Delta w = Q^T y$$

$$H = S^+ P^T = \left[\begin{array}{c} O_{n-r,m} \\ \hline T \end{array} \right], \qquad T \in R^{r \times m} \qquad (6.2.17)$$

$$L_n = I_n - S^+ S = \left[\begin{array}{c|c} I_{n-r} & O_{n-r,r} \\ \hline O_{r,n-r} & O_{r,r} \end{array} \right]$$

the inverse problem solution (6.2.16) can be presented in the form

$$\Delta v = H \Delta x_e + L_n \Delta w \qquad (6.2.18)$$

Due to the structure of matrices H and L_n, the above equation may be reduced to

$$\Delta v^u = \Delta w^u$$

$$\qquad (6.2.19)$$

$$\Delta v^\ell = T \Delta x_e$$

where $\Delta v = \left[\begin{array}{c} \Delta v^u \\ \hline \Delta v^\ell \end{array} \right]$, $\Delta v^u \in R^{n-r}$, $\Delta v^\ell \in R^r$, $\Delta w^u \in R^{n-r}$ are the upper and lower parts of the corresponding vectors. Equation (6.2.19) indicates that the coordinates of the generalized vector Δv are partially decoupled, with Δv^ℓ being completely dependent on the desired trajectory, while Δv^u may be chosen arbitrarily. Thus, the kinematic redundancy is concentrated in the first (n-r) components of the vector Δv.

Numerically stable algorithms and programs have been developed for evaluating the matrices P, Q and S [96]. However, the following question arises: what criterion should be used for choosing the (n-r) vector Δw^u. In [63] the necessary and sufficient conditions for the vector

Δw^u were derived, so that the joint angles stay within their constraints, $q_{imin} < q_i < q_{imax}$, $i=1,\ldots,n$.

The presence of kinematical constraints on joint coordinates

$$q_{imin} < q_i < q_{imax}, \qquad i=1,\ldots,n \qquad (6.2.20)$$

has also been considered in [83]. When one joint or more reaches its mechanical constraint, additional equations may be imposed, in order to prevent the violation of the joint coordinate constraints. The set of m linear equations (6.2.1) is then expanded by k equations of the form

$$\overset{(i)}{[0 \cdots 0\ 1\ 0 \cdots 0]}\Delta q = a_j \qquad j=m+1,\ldots,m+k, \quad m+k<n \qquad (6.2.21)$$

where k is the number of degrees of freedom which would violate the constraint if only Equation (6.2.1) would be satisfied. The numerical values of the constants a_j depend on the type of the constraint. Namely, for the constraints (6.2.20) a_j is set to zero, while for the constraints on joint velocities

$$|\dot{q}_i| < v_i, \qquad i=1,\ldots,n \qquad (6.2.22)$$

the constants a_j become

$$a_j = v_i\Delta t, \qquad \text{for} \quad \dot{q}_i > v_i$$
$$\qquad\qquad\qquad\qquad\qquad\qquad\qquad (6.2.23)$$
$$a_j = -v_i\Delta t, \qquad \text{for} \quad \dot{q}_i < -v_i$$

Equation (6.2.1) and (6.2.21) can be summarized into the matrix form

$$\Delta x_c = \begin{bmatrix} \Delta x_e \\ \hline a_{m+1} \\ \vdots \\ a_{m+k} \end{bmatrix} = J_c\Delta q = \begin{bmatrix} J \\ \hline A \end{bmatrix} \Delta q \qquad (6.2.24)$$

where $x_c \in R^{m+k}$ is the extended vector of the external coordinates' increment, $A \in R^{k \times n}$ is the matrix formed from the rows defined by (6.2.21), $J_c \in R^{(m+k) \times n}$ is the extended Jacobian matrix. The increment of joint coordinates Δq is still evaluated from (6.2.4) or (6.2.10), where J is

224

replaced by J_c and Δx_e by Δx_c.

Let us now consider the problem of the manipulator degeneracies, where the Jacobian matrix becomes singular. As we have seen, the evaluation of the matrix pseudoinverse according to the singular value decomposition, does not require full rank matrices. This evaluation is, however, considerably more computationally complex with respect to the evaluation of the generalized inverse (6.2.7), and requires about 15 times longer evaluation time [36]. Another way to overcome the singularity problem, is the same as with the nonredundant manipulators, i.e. to exclude the linearly dependent columns and rows from the Jacobian matrix, or to control the manipulator in joint coordinates.

Finally, it should be noticed that the optimal motion synthesis for redundant manipulators, discussed above, actually represents the sub-optimal motion synthesis from the standpoint of the entire trajectory, since the minimization of the optimality criterion is carried out on short time intervals Δt, which correspond to the small increments Δx_e from the system (6.2.1). This practically implies the following. If, for example, the minimum norm solution (6.2.7) is considered for a manipulator moving along a closed spatial path (e.g. a square, Fig. 6.1),

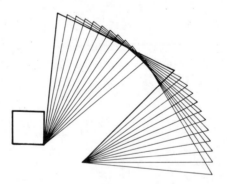

Fig. 6.1. Three degree-of-freedom manipulator
moving along a square path

the procedure for the optimal evaluation of joint coordinates increment, need not yield, in a general case, the same values for the joint coordinates upon return to a given point of the closed path. Depending on the manipulator kinematic structure and the distance between the path and the manipulator base, the joint coordinates drift an almost constant amount each cycle, otherwise there is a bounded region in which they change [88]. By imposing additional constraints of the type (6.2.21), one can provide that the joint angles stay within the allowable region. The drift of angles can also be prevented by introducing the homogeneous term which would, for example, minimize the criterion (6.2.12).

The methods presented in this section represent the main results achieved in solving the inverse kinematic problem for redundant manipulators, in the absence of obstacles. The manipulator end-effector trajectory synthesis in the external coordinate space, and the tracking of these trajectories may be carried out in the same way as for nonredundant manipulators.

6.3. Energy Optimal Motion Synthesis

The subject of interest in this section will be the dynamic approach to the nominal trajectory synthesis for redundant manipulators. Most of the methods proposed in the literature are concerned with the kinematic approach, and little work has been done in the domain of dynamic motion synthesis for redundant manipulators [83], although the dynamic properties of the system play an important role in the behaviour of the overall system. The control synthesis should, therefore, take the dynamic effects into account, particularly when high speed motion or the manipulation of heavy objects by means of powerful robots are considered.

In this section we will present a procedure for the dynamic synthesis of redundant manipulator trajectories [97-98]. This procedure (like the one presented in [83]), is not really dynamic, in the sense defined in Section 5.3, for the reason that the system is modelled by the kinematic model, but the optimality criterion is a dynamic one. The performance index is the energy consumed by the actuators during the motion, evaluated from the dynamic model of the mechanism and the actuators. Here, as with the kinematic methods for redundant manipulator motion synthesis, we are actually dealing with partial optimal motion synthesis, on the time intervals Δt. The trajectory obtained is generally suboptimal with respect to that which would be obtained by minimization on the complete motion execution time [0, T].

We will consider the problem of distributing the motion among the degrees of freedom, i.e. the solution of the inverse kinematic problem, given a manipulator end-effector trajectory in the external coordinate space, $x_e(t) \in X_e \in R^m$. The solution to the underdetermined system of linear equations

$$\dot{x}_e = J(q)\dot{q} \qquad (6.3.1)$$

is to be found, where $J \in R^{m \times n}$, $m < n$, m is the number of external coordinates specifying the task, n is the number of degrees of freedom.

The optimality criterion introduced in [83] to eliminate the indefiniteness of the solution, was the kinetic energy of the mechanism

$$\Omega(\dot{q}) = \dot{q}^T H(q)\dot{q} \qquad (6.3.2)$$

where $H(q) \in R^{n \times n}$ is the positive definite matrix of kinetic energy, i.e. the inertial matrix $H(q)$ in the dynamic model (5.2.2). In [83] the matrix was formed analytically for a 4 degree-of-freedom manipulator, and the joint trajectories obtained for a given Cartesian trajectory $x(t)$, $y(t)$, $z(t)$, $t \in [0, T]$, have been shown.

In the procedure we are about to describe here, the idea is to take the entire energy consumed by the actuators as the optimality criterion. In this case, the complete model of the system (the model of the mechanism and the actuators) should be taken into account. We will consider the two types of actuators, most frequently encountered with industrial robots: electric DC motors and hydraulic actuators. The special case, when the hydraulic actuator is nonlinearly coupled with its degree of freedom, so that a nonlinear relationship exists between the actuator coordinate (linear displacement of the cylinder) and the joint coordinate (joint angle), is also considered.

For these two types of actuators the energy consumption can be presented in the form

$$\Omega(\dot{q}) = \frac{1}{2}\dot{q}^T M_1 \dot{q} + M_2^T \dot{q} \qquad (6.3.3)$$

where $M_1 \in R^{n \times n}$ is a symmetric positive definite matrix, $M_2 \in R^n$. The above expression will be derived later in this section. The optimal velocity $\dot{q}$ is obtained, as explained in Section 6.2, by considering the Lagrangian

$$L(\dot{q}, p) = \frac{1}{2} \dot{q}^T M_1 \dot{q} + M_2^T \dot{q} + p^T(\dot{x}_e - J\dot{q})$$

By imposing the condition that the partial derivatives of the Lagrangian with respect to $\dot{q}$ and the adjoint vector p, have to be zero at the optimal point, the solution is obtained in the form

$$\dot{q} = G \dot{x}_e + (GJ - I_n) M_1^{-1} M_2 \qquad (6.3.4)$$

where G is the generalized inverse of J

$$G = M_1^{-1} J^T (J M_1^{-1} J^T)^{-1} \qquad (6.3.5)$$

and I_n is the n×n identity matrix. The kinematic constraints (6.2.20) – (6.2.24) can also be introduced into the motion synthesis.

The matrices M_1 and M_2 depend on the joint coordinates $q(t_k)$, i.e. the time instant t_k at which the optimization on the interval $\Delta t = (t_k, t_{k+1})$, $k = 0, \ldots, N-1$, $T = N\Delta t$ is carried out. Therefore, the cost functional (6.3.3) may be more precisely written in the form

$$\Omega(\dot{q}^{(k)}) = \frac{1}{2} \dot{q}^{(k)T} M_1(t_k) \dot{q}^{(k)} + M_2^T(t_k) \dot{q}^{(k)} \qquad (6.3.6)$$

where $\dot{q}^{(k)} = \dot{q}(t_k)$, $M_1(t_k) \in R^{n \times n}$, $M_2(t_k) \in R^n$.

Let us now discuss the evaluation of these matrices for electric and hydraulic actuators.

Energy consumption of electric DC motors

As we have said, we will consider the case when the manipulator joints are driven by electric DC motors. We will consider only the case when there exists a linear relationship between the actuator coordinates q_a and the joint coordinates q (i.e. $q = q_a$), since the nonlinear relationship is not encountered in practice. The second order model of DC motors is given by (5.2.4), (5.2.8) – (5.2.9) in Section 5.2.

The power delivered to the ith motor is

$$W^i = u_i i_i = d_i P_{Mi}^2 + c_i P_{Mi} \dot{q}_i \qquad (6.3.7)$$

where

$$d_i = R_i/(K_{emi}N_{mi})^2, \quad c_i = \frac{K_{mei}N_{vi}}{K_{emi}N_{mi}} \tag{6.3.8}$$

$$P_{Mi} = P_i + J_i\ddot{q}_i + F_i\dot{q}_i \tag{6.3.9}$$

The parameters R_i, K_{emi}, K_{mei}, N_{mi}, N_{vi}, J_i and F_i have already been defined in Section 5.2. P_i denotes the driving torque (force) action on the motor output shaft. It is determined by the manipulator dynamic model

$$P = H(q)\ddot{q} + h(q, \dot{q}) \tag{6.3.10}$$

where $P = [P_1 \cdots P_n]^T$, $H \in R^{n \times n}$ - is the inertial matrix, $h(q, \dot{q}) \in R^n$ is the vector of the gravitational, centrifugal and Coriolis forces. Friction in the joints may be included into the parameter F_i, which specifies the friction in the actuator reducer.

The energy consumption of the complete system on the time interval (t_k, t_{k+1}) is approximately evaluated as $\Delta E(t_k) = W(t_k)\Delta t$. It can be presented in the following matrix form

$$\Delta E^{(k)} = (P_M^{(k)T} D\ P_M^{(k)} + P_M^{(k)}C\ \dot{q}^{(k)})\Delta t \tag{6.3.11}$$

where $P_M = [P_{M1} \cdots P_{Mn}]^T$, $D \in R^{n \times n}$, $D = \text{diag}(d_i)$, $C \in R^{n \times n}$, $C = \text{diag}(c_i)$. The torques (forces) produced by the motors are

$$P_M(t_k) = P_M^{(k)} = P(t_k) + J_M\ddot{q}^{(k)} + F_M\dot{q}^{(k)} \tag{6.3.12}$$

where $J_M \in R^{n \times n}$, $J_M = \text{diag}(J_i)$, $F_M \in R^{n \times n}$, $F_M = \text{diag}(F_i)$. By substituting (6.3.10) into (6.3.12) and adopting $\ddot{q}^{(k)} \cong (\dot{q}^{(k)} - \dot{q}^{(k-1)})/\Delta t$ and $h^{(k)} = h(q^{(k)}, \dot{q}^{(k-1)})$, we obtain

$$P_M^{(k)} = H_1^{(k)}\dot{q}^{(k)} + \frac{1}{\Delta t} H_2^{(k)} \tag{6.3.13}$$

where

$$H_1^{(k)} = \frac{1}{\Delta t}(H^{(k)} + J_M) + F_M$$
$$H_2^{(k)} = h^{(k)}\Delta t - (H^{(k)} + J_M)\dot{q}^{(k-1)} \tag{6.3.14}$$

By substituting (6.3.13) into (6.3.11) one obtains the energy consumption in the form

$$\Delta E^{(k)} = \frac{1}{2} \dot{q}^{(k)T} M_1(t_k) \dot{q}^{(k)} + M_2^T(t_k) \dot{q}^{(k)} + M_3(t_k) \qquad (6.3.15)$$

where

$$M_1(t_k) = 2\Delta t \; H_1^{(k)}(DH_1^{(k)}+C)$$

$$M_2(t_k) = (2 \; H_1^{(k)}D+C)H_2^{(k)} \qquad (6.3.16)$$

$$M_3(t_k) = \frac{1}{\Delta t} H_2^{T(k)} D \; H_2^{(k)}$$

It is evident that the first two terms in Equation (6.3.15) also exist in (6.3.6), while the term M_3 does not depend on $\dot{q}^{(k)}$, and therefore can not be included into the optimization.

Energy consumption of hydraulic actuators

We will consider the following two cases encountered in practice for the hydraulic manipulators:

1. a linear relationship exists between the *i*th actuator coordinate q_{ai} (the linear displacement of the piston or the angle of rotation for vane hydraulic actuators) and the corresponding joint coordinate q_i (joint angles or linear displacements for sliding joints);

2. a nonlinear relationship axists between the actuator coordinate q_{ai} and the joint coordinate q_i.

In industrial practice, the relationship is usually linear, but the second case also sometimes appears. The example for such a degree of freedom is the second joint of the UMS-3B manipulator shown in Fig. 6.5, where the hydraulic cylinder drives the revolute joint. In practice, the coordinates q_{ai} and q_i most frequently coincide. On the other hand, the case when the relationship is linear can easily be derived from the nonlinear case. So, we will here consider the case when these coordinates coincide together with the general nonlinear relationship.

Let us now consider the evaluation of the matrices M_1 and M_2 from the optimality criterion (6.2.6). The power produced by the *i*th hydraulic cylinder, when it is loaded by external force F_{ai}, may be presented as

$$W_i = (m_i \ddot{q}_{ai} + B_{ci} \dot{q}_{ai} + F_{ai}) \dot{q}_i \qquad (6.3.17)$$

where q_{ai} is the actuator coordinate (the linear displacement of the piston), m_i - the piston mass (together with the rod), B_{ci} - the viscous friction coefficient. For the rotational hydraulic actuator an analogous expression can be written. The total power produced by all the actuators can be presented in the matrix form

$$W = \ddot{q}_a^T M \ddot{q}_a + \dot{q}_a^T B_c \dot{q}_a + \dot{q}_a^T F_a \qquad (6.3.18)$$

where $M \in R^{n \times n}$, $M = \text{diag}(m_i)$, $B_c \in R^{n \times n}$, $B_c = \text{diag}(B_{ci})$, $F_a \in R^n$ is the vector of generalized driving forces and $q_a \in R^n$ is the actuator coordinate vector. We will first assume that the joint and actuator coordinates coincide, $q_a = q$. In this case the generalized driving forces are determined directly by the dynamic model of the mechanism (6.3.10), $F_a = P$.

By substituting (6.3.10) into (6.3.18) and introducing the same approximation for $h(q, \dot{q})$ as for the electric actuators, the energy consumed on the time interval (t_k, t_{k+1}) is obtained in the form (6.3.6), with the matrices M_1 and M_2 being

$$M_1(t_k) = 2(M + B_c \Delta t + H^{(k)})$$

$$M_2(t_k) = h^{(k)} \Delta t - (M + H^{(k)}) \dot{q}^{(k-1)} \qquad (6.3.19)$$

Let us now prove that the matrix M_1 is symmetric and positive definite. Since the matrices M and B_c are diagonal and the inertial matrix H is symmetric [16, 65], it follows directly from Equation (6.3.19) that M_1 is also symmetric. Further, as the inertial matix H is positive definite, and matrices M and B_c are diagonal with positive elements, i.e. they are positive definite, the matrix M_1 (being the linear combination of these matrices) is also positive definite. Therefore, the inner product $\langle x, M_1 x \rangle = x^T M_1 x$ is positive for all $x \in R^n \neq 0$.

Let us now consider the general case when the actuator coordinates are nonlinearly related to the mechanical joint coordinates

$$q_{ai} = g_i(q_i), \qquad i = 1, \ldots, n \qquad (6.3.20)$$

Then the generalized driving forces F_a are related to the generalized torques (forces) P acting at the joints by

$$F_a = C_F P \tag{6.3.21}$$

where $C_F \in R^{n \times n}$, $C_F = \text{diag}(g_i^{-1}(q_i))$, and P is given by (6.3.10). The inverse functions g_i^{-1} are usually available in analytical form. The relationship between the rates and accelerations is given by

$$\dot{q}_a = G_1 \dot{q}, \qquad \ddot{q}_a = (G_2 \dot{q}) \dot{q} + G_1 \ddot{q} \tag{6.3.22}$$

where $G_1 \in R^{n \times n}$, $G_1 = \text{diag}(\partial g_i / \partial q_i)$, $G_2 \in R^{n \times n \times n}$, $G_2 = \text{diag}(\partial^2 g_i / \partial q_i^2)$. Inserting these relations into (6.3.18) one obtains the energy as a cubic function of $\dot{q}$. In order to reduce it to the form (6.3.6) we have to introduce the approximation $\dot{q}^{(k)} \cong \dot{q}^{(k-1)}$. Then we obtain the matrices of the performance index (6.3.6) as

$$M_1(t_k) = 2[G_1^{(k)} M(G_2^{(k)} \dot{q}^{(k-1)} \Delta t + G_1^{(k)}) + G_1^{(k)} B_c G_1^{(k)} \Delta t + H^{(k)}]$$

$$\tag{6.3.23}$$

$$M_2(t_k) = h^{(k)} \Delta t - (G_1^{(k)} M G_1^{(k)} + H^{(k)}) \dot{q}^{(k-1)}$$

If the relationship (6.3.20) is linear, $q_{ai} = C_{1i} q_i + C_{oi}$, the matrix G_2 becomes a zero matrix, and the approximation of $\dot{q}^{(k)}$ mentioned is not necessary. The matrix G_1 becomes a constant matrix, so that $G_1 M G_1$ and $G_1 B_c G_1 \Delta t$ matrices may be evaluated in advance.

Although these expressions seem rather complex, their evaluation simplifies a great deal for real, industrial manipulators, since the number of degrees of freedom being nonlinearly coupled with the actuators is usually very small (one, or two). The program implementation of the algorithm should certainly take into account the particular structure of matrices M, B_c, H (diagonality and symmetry).

The energy optimal motion synthesis can be summarized as follows: at each integration interval the Jacobian matrix $J(q)$, the inertial matrix $H(q)$ and the vector of gravitational and centrifugal forces $h(q, \dot{q})$ should be evaluated, given the trajectory of manipulator hand $x_e(t)$; then, matrices M_1 and M_2 are calculated, as well as the inverse of M_1 and $J M_1^{-1} J^T$; then the generalized inverse is evaluated according to (6.3.5) and the optimal value of q according to (6.3.4). In order to speed up the computation, the analytical kinematic and dynamic models should be employed [16-18].

Numerical examples

In order to illustrate the above algorithm we will first consider the electrically driven, 6 degree-of-freedom manipulator shown in Fig. 6.2.

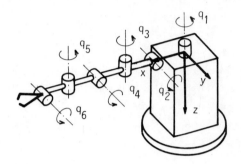

Link	1	2	3	4	5	6
Mass [kg]	−	2.45	0.67	0.46	0.46	2.1
Length [m]	0.14	0.35	0.26	0.1	0.1	0.15
J_N [kgm^2]	−	0.035	0.0033	0.0006	0.0006	0.001
J_S [kgm^2]	0.16	0.004	0.0007	0.0005	0.0005	0.0007

Fig. 6.2. Redundant electrical manipulator

The dynamic parameters of the manipulator are shown in the table in Fig. 6.2, while the actuator parameters (6.3.8) and actuator moments of inertia J_i and friction coefficients F_i are shown in Table T.6.1.

Link	1,2,4,5,6	3
c_i [MKSA]	1.87	2.48
d_i [MKSA]	0.13	0.34
J_i [kgm^2]	1.52	3.3
F_i [N/m/s]	1.23	1.4

Table T.6.1. The actuator parameters of the
electrical manipulator

The end-effector trajectory is specified by three Cartesian coordinates, $x_e = [x\ y\ z]^T$ in the base reference coordinate frame Oxyz. The straight line motion between an initial point x_e^O and the final point x_e^F is desired

$$x_e(t) = x_e^O + \lambda(t)(x_e^F - x_e^O), \qquad 0 < t < T \qquad (6.3.24)$$

The parameter $\lambda \in [0, 1]$ specifies the velocity distribution along the trajectory. Here, we will take

$$\dot\lambda(t) = (1 - \cos(2\Pi t/T))/T$$

where T is the motion execution time.

In the example to follow the movement between the points $x_e^O = [1\ 0.15\ 0.28]^T$[m] and $x_e^F = [0.2\ 0.94\ 0.3]^T$ [m] in T = 1.8 s will be considered. The optimal joint trajectories obtained are presented by solid lines in Fig. 6.3. The time history of joint coordinates obtained for the performance index being $\frac{1}{2}\Delta q^T I_n \Delta q$, are presented by dashed lines in the same figure. The time history of the energy consumption increment ΔE, as well as the total energy consumed are shown in Fig. 6.4. It is evident that the optimization with criterion being the real energy consumption yields much better results than the minimum norm solution. The energy performance index penalizes more those degrees of freedom that consume more energy, thus favoring fast motion of those links that are less loaded. These are usually the end-effector joints (e.g. q_6 in Fig. 6.3).

The second example is the hydraulically driven UMS3B manipulator shown in Fig. 6.5. Its dynamic parameters are shown in the table in Fig. 6.5. The actuator parameters relevant for the energy calculation are given in Table T.6.2. As we have already mentioned, there exists a nonlinear relationship between the joint angles q_1 and q_2 and the linear displacement of actuators q_{a1} and q_{a2}, since the linear motion of the pistons is transformed into the rotational motion of the first two degrees of freedom. In this example, the nonlinear relationship (6.3.20) is described by the following relations

$$q_{a1} = 0.12\ tg\ q_1, \qquad \dot q_{a1} = 0.12\ \dot q_1/\cos^2 q_1$$

$$q_{a2} = (1.19^2 + 0.2^2 - 2 \cdot 0.2 \cdot 1.19\ \cos(q_2 + 1.4))^{1/2}$$

$$\dot q_{a2} = 2\dot q_2 \sin(q_2 + 1.4) \cdot 0.2 \cdot 1.19/q_{a2}$$

The manipulator end-effector path is specified by three Cartesian co-

ordinates, so that there exist three degrees of freedom more than it is
required. The motion between the points x_e^O = [0 1.28 0.94]^T [m]
x_e^F = [-1. 0.5 1.5]^T [m], T = 2 s is desired. The energy optimal distribution of joint coordinates and the distribution obtained by mini-
mizing $\frac{1}{2}\Delta q^T I_n \Delta q$ are presented in Fig. 6.6. The corresponding energy
consumption distributions are presented in Fig. 6.7.

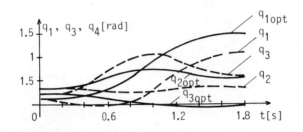

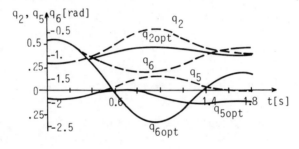

Fig. 6.3. Optimal joint coordinates

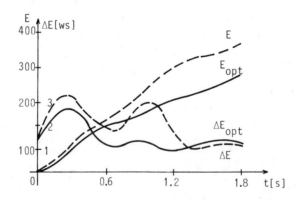

Fig. 6.4. Time history of energy consumption

Comparing these examples one could conclude that the energy savings for
hydraulic manipulator are higher than that for electrically powered
robots. However, this conclusion should be taken with certain reserva-
tions, since the energy evaluations are not quite consistent. The power

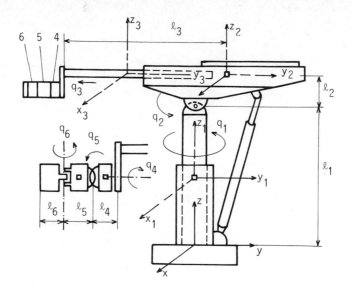

Link	1	2	3	4, 5	6
Mass m_i [kg]	–	27.8	22.34	2.33	3.33
Length ℓ_i [m]	1.2	0.142	1.14	0.14	0.26
J_{ix} [kgm^2]	0	2.98	1.21	0.004	0.007
J_{iy} [kgm^2]	0	–	–	0.004	0.009
J_{iz} [kgm^2]	0.322	3.701	1.21	0.004	0.009

Fig. 6.5. The UMS-3B hydraulic manipulator

Actuator	1, 2	3	4, 5, 6
m_i	2.65 [kg]	3.07 [kg]	0.003 [kgm^2]
B_{ci}	30. $\left[\dfrac{N}{m/s}\right]$	30. $\left[\dfrac{N}{m/s}\right]$	0.5 $\left[\dfrac{Nm}{rad/s}\right]$

Table T.6.2. Actuator parameters for the
UMS-3B manipulator

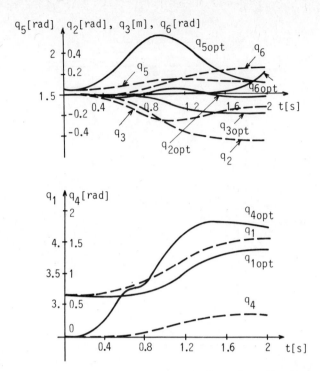

Fig. 6.6. Optimal joint trajectories for the UMS-3B manipulator

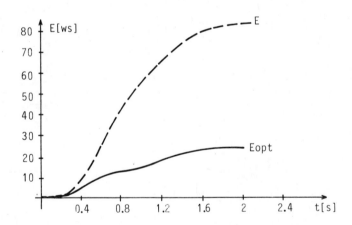

Fig. 6.7. Time history of the energy consumption

used in the hydraulic power calculations is the power delivered to the piston of the actuator. The piston mass, viscous friction and the load are taken into account. The power losses in the hydraulic power supply and the lines are not included. On the other hand, the power used in the electric power calculations is the power delivered to the motor from the electric power supply. Therefore, these two examples are not directly comparable.

The procedure presented is primarily aimed towards off-line nominal trajectory synthesis, since it requires the calculation of the dynamic model matrices. However, the development of real-time dynamic modelling techniques and the increase in computational speed of modern computers offer a possibility for real-time energy optimal trajectory synthesis.

Within this approach to redundant manipulator motion synthesis, the dynamic properties of the system are taken into consideration by means of the optimality criterion. This method exhibits considerable advantages over kinematic approaches in the following cases: manipulation of heavy objects by large, powerful robots and high speed manipulation with high energy consumptions. Significant energy savings are attained by increased computational complexity with respect to purely kinematic approaches.

6.4. Obstacle Avoidance Using Redundant Manipulators

As we have already mentioned in the introduction to this chapter, the obstacle avoidance problem is an important investigation task nowadays. It is a complex problem involving several subproblems. In [53-56] the problem of searching for collision-free paths, given an initial and final location and orientation of a work object and a set of obstacles located in space, has been considered. This problem is not closely related to redundant manipulators, and we are not going to discuss it in this book.

In this section we will be concerned with finding such joint trajectories which provide for collision-free motion of all the manipulator links, given an end-effector trajectory, a set of obstacles in space and an initial manipulator configuration. We will here present the main results achieved so far in this domain [81, 83, 89, 99-103].

Let us consider a serial link manipulator with n degrees of freedom. The joint coordinate vector $q = [q_1 \cdots q_n]^T$ belongs to the configuration space $Q = \{q: q_{imin} < q_i < q_{imax}, i=1,\ldots,n\}$, where q_{imin} and q_{imax} denote the minimum and maximum values of the ith joint coordinate, defined by the mechanical constraints of the mechanism. Each point in the manipulator work space may be specified by a position vector $x = [x\ y\ z]^T \in R^3$, where x, y, z are the Cartesian coordinates with respect to the reference coordinate frame $Oxyz$ (Fig. 6.8).

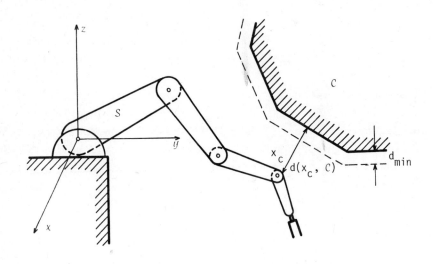

Fig. 6.8. Manipulator work space

Let us denote by S the subset of points in the work space which are occupied by the manipulator links. This subset is uniquely defined by the manipulator configuration, i.e. by the vector of joint coordinates q. Let us assume that in the manipulator environment there exists a convex body representing an obstacle. The subset of points occupied by the obstacle will be denoted by C.

We will assume that the manipulator initial configuration is given $q(0) = q^o$ and the corresponding external coordinate vector $x_e^o = f(q^o) \in R^m$, $m < n$. The manipulator end-effector motion will be specified either by a trajectory $x_e(t)$, $t \in [0, T]$, or by a set of points x_e^k, $k=0,\ldots,N$ which the manipulator should pass through.

The problem is to determine a trajectory $q(t)$ so that the given trajectory $x_e(t)$ (or the points x_e^k) is achieved and the collisions avoided. It will be considered that the manipulator has avoided a convex obstacle C, if the following condition is satisfied

$$d(S,\ C) > d_{min}, \qquad \forall t \in [0,\ T] \tag{6.4.1}$$

where

$$d(S, C) = \min_{x \in S} d(x, C) \tag{6.4.2}$$

represents the minimal distance between the manipulator and the obstacle, and d_{min} is a given minimum value of this distance. Here

$$d(x, C) = \min_{y \in C} ||x-y|| \tag{6.4.3}$$

denotes the distance between the point x on the arm and the obstacle C, $||x||$ is a given norm in the space R^n.

The problem of evaluating the minimal distance between the obstacle and the arm is very complex. It will not be considered in this book. Its solution should employ the obstacle modelling techniques, information obtained from sensors, etc. We will here assume that the minimal distance is evaluated at a higher control level, where all ambiguous situations (when two or more points on the arm are at the same distance from the obstacles) are resolved, too.

The above-stated problem is inherently related to redundant manipulators. It involves the inverse kinematic problem solution, i.e. solving the set of equations

$$\Delta x_e = J(q) \Delta q \tag{6.4.4}$$

where $\Delta x_e \in R^m$, $\Delta q \in R^n$, $J \in R^{m \times n}$, $m < n$. The general solution is given by (6.2.10), i.e. by

$$\Delta q = G_1 \Delta x_e + (G_2 J - I_n) y \tag{6.4.5}$$

where G_1 and G_2 are the generalized inverse matrices of J, satisfying $JG_1J = J$, $JG_2J = J$, $y \in R^n$ is an arbitrary vector. As we have discussed in Section 6.2, if the vector y is chosen as $y = \alpha \nabla H(q)$, then the solution (6.4.5) forces the function $H(q)$ to decrease.

In [81, 89] the function $H(q)$ was adopted in the form (6.2.12), i.e. as

$$H(q) = \frac{1}{2}(q-q_r)^T H_1 (q-q_r) \tag{6.4.6}$$

where $H_1 \in R^{n \times n}$, $H_1 = \text{diag}(h_{1i})$, h_{1i} are positive constants, and $q_r \in R^n$ a given reference joint coordinate vector. The inverse problem solution then becomes

$$\Delta q = G_1 \Delta x_e + \alpha (G_2 J - I_n) H_1 (q - q_r) \tag{6.4.7}$$

The second term on the right-hand side provides that the joint coordinates stay as close as possible to the given reference values. However, the evaluation of the reference vector q_r, depending on the disposition of obstacles and the manipulator kinematic structure is not simple, due to the highly nonlinear relationship between the arm position and the joint coordinates and the ambiguity of the solution.

In [103] the problem considered is the evaluation of such joint trajectories which ensure that the inboard links stay away from the obstacles while approximately guiding the hand on the disered trajectory. The concept of pseudoinverse matrices is used. The solution (6.4.5) is presented in the form

$$\dot{q} = J^+ \dot{x}_e + Py$$

$$P = I_n - J^+ J \tag{6.4.8}$$

where J is the Jacobian matrix for the end-effector and J^+ is its generalized inverse. The arbitrary vector y will be chosen in the following way. The velocity of the manipulator point being closest to the obstacle is set to be $\dot{x}_c$, having the direction away from the obstacle

$$J_c \dot{q} = \dot{x}_c \tag{6.4.9}$$

where $J_c \in R^{3 \times n}$ is the Jacobian matix which corresponds to the critical point on the arm. Since the critical point belongs to any link ℓ, $\ell = 1, \ldots, n-1$, the matrix J_c depends only on the first ℓ coordinates, i.e. the last $(n-\ell)$ columns of $J_c \in R^{3 \times n}$ are zero columns.

By substituting (6.4.8) into (6.4.9) one obtains the equation

$$J_c Py = \dot{x}_c - J_c J^+ \dot{x}_e \tag{6.4.10}$$

Since the matrix $J_c P$ is singular, its pseudoinverse matrix is proposed in order to obtain the least square fit for the vector y. This result is then substituted back into (6.4.8) to solve for joint velocities. Upon simplification, the solution is proposed in the form

$$\dot{q} = J^+\dot{x}_e + [J_cP]^+[\dot{x}_c - J_cJ^+\dot{x}_e] \qquad (6.4.11)$$

It is evident, however, that the second term on the right-hand side is no more a homogeneous term, so that (6.4.11) is no longer the exact solution to the system (6.4.4). This means that the tracking of the desired trajectory is not exactly provided, although there might exist a feasible solution, without deviating from the given trajectory.

The change of the point closest to the obstacle changes the rank of the matrix J_c, resulting in a discontinuity in joint velocities. In order to minimize the rank change effects the solution for the joint rates has been modified to

$$\dot{q} = J^+\dot{x}_e + \alpha_h[J_cP]^+(\alpha_c\dot{x}_c - J_cJ^+\dot{x}_e) \qquad (6.4.12)$$

where the scalar parameters α_c and α_h are chosen to depend on the distance from the obstacle. A second way of minimizing the switching of homogeneous solutions is to express $\dot{q}$ as

$$\dot{q} = J^+\dot{x}_e + \alpha_1(d_2/d_1)h_1 + \alpha_2(d_2/d_1)h_2 \qquad (6.4.13)$$

where h_1 and h_2 are two homogeneous solutions for two critical points, and d_1 and d_2 are the distances to the obstacle. The greater the disparity between d_1 and d_2, the closer α_1 comes to unity, with α_2 approaching zero. With d_1 approximately equal d_2, $\alpha_1 = \alpha_2 = 0.5$.

A completely different approach to the obstacle avoidance problem by means of redundant manipulators, was presented in [83, 100]. The problem considered was the evaluation of such joint trajectory which provides for the manipulator hand motion towards a final point, together with preventing the manipulator links from approaching the obstacles. The manipulator moves in a field of forces, where the final point is an attractive pole for the gripper, while the obstacles are the repulsive surfaces. A potential function V_a, being concave and positive, is introduced in order to provide motion towards .the final point x_e^F

$$V_a = V_a(x_{e1}, \ldots, x_{em}) > 0$$

$$\qquad (6.4.14)$$

$$V_a = 0 \iff x_e = x_e^F$$

The attractive pole affects the manipulator motion through the generalized forces F_a, which act directly on the manipulator joints (beside the gravitational forces F_g and dissipation term F_d)

$$F_a = J^T(q) \, \text{grad} \, V_a \tag{6.4.15}$$

Similarly, a potential function V_o is introduced, with the purpose of repulsing the manipulator from the obstacles. An obstacle - a body θ_i in the work space is modelled by the following set of equations E_i

$$E_i \triangleq \{f_i^j(x, y, z) = 0; \, j=1, \, n_i\} \tag{6.4.16}$$

where $f_i^j(x, y, z) = 0$ are the analytical equations associated with simple geometric objects. For example, the parallelepiped is described by

$$(\frac{x-x_o}{a})^8 + (\frac{y-y_o}{b})^8 + (\frac{z-z_o}{c})^8 = 1 \tag{6.4.17}$$

Similarly, the equations describing a cone and a cylinder are given. The potential function V_i^j, is introduced for each function f_i^j, having the following properties: V_i^j is a nonnegative function, with a continuous gradient; it rises towards infinity at the surface of the obstacle. So, the V_i^j function was chosen as

$$V_i^j(x, y, z) = (1/f_i^j(x, y, z) - 1/f_i^j(x^*, y^*, z^*))^2,$$
$$\text{for } |f_i^j(x, y, z)| < |f_i^j(x^*, y^*, z^*)| \tag{6.4.18}$$

$$V_i^j(x, y, z) = 0, \qquad \text{for } |f_i^j(x, y, z)| > |f_i^j(x^*, y^*, z^*)|$$

where x^*, y^*, z^* denotes a given point in the work space. The total potential function V_o^i for the obstacle θ_i is obtained from

$$V_o^i = \sum_{j=1}^{n_i} V_i^j(x, y, z) \tag{6.4.19}$$

Rejection is now provided by adding the term F_o to the generalized forces acting at the manipulator joints

$$F_o^i = J^T(q) \, \text{grad} \, V_o^i(x, y, z) \tag{6.4.20}$$

If there are several objects θ_i representing the obstacles, the total rejecting force is $F_o = \sum F_o^i$. The total vector of generalized forces (torques) acting in the joints is now evaluated as

$$F = F_d + F_g + F_a + F_o \qquad (6.4.21)$$

where F_a is given by (6.4.15), F_d and F_g are the dissipative and gravitational terms.

We may conclude from the above discussion that the motion of the end-
-effector is not strictly defined. What motion will actually be realized, and whether the end point will be reached or not, depends also
on the dynamic parameters of the mechanism and the actuators and the
parameters of the potential functions introduced. The problems arising
in the implementation of the above algorithm are concerned with the
evaluation complexity of these functions and their gradients, the discontinuity of the gradients at the intersection lines, the stability
of the solution.

The optimal motion of redundant manipulators in environments with obstacles was also considered in [101]. This approach is based on the
linear programming optimization technique. The problem considered involves the evaluation of such a set of admissible joint coordinate vectors (configurations) $q(1) = q^o$, $q(2),\dots,q(k),\dots,q(N)$, which transfers
the end-effector from the initial state x_e^o to the final state x_e^F. The
admissible configuration q lies within the physical constraints and
satisfies the nonequality (6.4.1), with the obstacle C being a convex
body. The joint trajectory should be determined in such a way that the
end-effector trajectory deviation from the straight line segment $x_e^F - x_e^o$,
is minimized.

The stated problem was reduced to application of the linear programming
technique in the following way. The property of the distance function
$d(x, C)$ (6.4.3), that it is a convex function, is utilized. It satisfies the subgradient equation

$$d(z, C) > d(x, C) + <\nabla d, z-x> \qquad (6.4.22)$$

for any $z \in R^n$, $<x, y> = \sum_{i=1}^{n} x_i y_i$. The vector ∇d is termed the subgradient
of the convex function $d(x, C)$ in the point x (a generalization of the
gradient for the functions which are not smooth [104]). The set of all
the subgradients of convex function $f(x)$ is denoted by $\partial f(x)$, i.e. in
this case by $\partial d(x, C)$. If $f(x)$ is a differentiable in the point x, the
set consist of a single vector, this being $\partial f/\partial x$.

Let us assume that at the kth step in trajectory planning, the distance between a link ℓ and the obstacle C is given by $d(x_\ell[k], C)$, where $x_\ell[k]$ denotes the point on the ℓth link which is closest to the obstacle. The evaluation of the increment of joint coordinates $\Delta q[k]$ at the kth step, should provide that the displacement $\Delta x_\ell[k]$ of the critical point satisfies the nonequality

$$d(x_\ell[k], C)+(\nabla d_\ell^k, \Delta x_\ell[k]) > d_{min} \qquad (6.4.23)$$

where ∇d_ℓ^k is the subgradient of the function $d(x, C)$ at the point $x_\ell[k]$. Assuming that the increments are small, the relationship $\Delta x_\ell[k] = J_\ell[k]\Delta q[k]$ may be established, with $J_\ell[k]$ being the Jacobian matrix for the critical point $x_\ell[k]$. The nonequality (6.4.23) can now be rewritten in the equivalent form

$$-\langle J_\ell^T[k]\nabla d_\ell^k, \Delta q[k] \rangle < d(x_\ell[k], C)-d_{min}, \qquad \ell=1,\ldots,n \qquad (6.4.24)$$

The constraints originating from the presense of obstacles have thus been expressed as the linear constraints with respect to the increment of joint coordinates $\Delta q[k]$. At each optimization step k, the distances from the obstacles to all the links are to be determined, as well as the subgradients ∇d_ℓ^k, and the Jacobian matrices $J_\ell[k]$ for the critical points on the links. The complexity of this problem depends a great deal on the definition of the distance function and the class of objects considered as obstacles. In [101] the distance function was adopted to be the Chebishev norm $||x||_\infty = \max\limits_{i=1,\ldots,n} |x_i|$, the manipulator is represented by broken lines, while the obstacles are convex polyhedral bodies. These assumption make the use of the linear programming possible.

The increment of joint coordinates $\Delta q[k]$ is evaluated optimally with respect to the criterion

$$\min_{\Delta q[k]} \{ \sum_{i=1}^{n} \gamma_i |\Delta q_i[k]| + \max_{j=1,\ldots,n} |(\Delta x_e[k]-J[k]\Delta q[k])_j| \} \qquad (6.4.25)$$

given the straight line segment $x_e^F-x_e^O$ in the external coordinate space, the constraints (6.4.24) and $q_{imin} < q_i < q_{imax}$, $i=1,\ldots,n$. Here the $\Delta x_e[k] = x_e[k+1]-x_e[k] \in R^m$ represents the desired increment of external coordinates, $J[k] \in R^{m \times n}$ is the Jacobian matrix relating between the external velocities and the joint velocities, γ_i - are scalar coefficients. The optimality criterion (6.4.25) represents a weighted sum of the joint increments plus the error in the Chebishev sense of the solution to the system $\Delta x_e[k] = J[k]\Delta q[k]$. The optimization task (6.4.25), constrained by (6.4.24) and physical limitations on joint angles, is reduced to the standard linear programming task

$$\min \ c^T x, \quad Ax \leqslant B, \quad x \geqslant 0 \qquad\qquad (6.4.26)$$

upon introducing the following notations: $\Delta q_i[k] = x_i - x_{n+i}$, $x_i, x_{n+i} \geqslant 0$, $x_{2n+1} = \max\limits_{j=1,\ldots,n} |\Delta x_e[k] - J[k]\Delta q[k]| \geqslant 0$, $x = [x_1 \ x_2 \ \cdots \ x_{2n+1}]^T$, $c^T = [\gamma^T \vdots \gamma^T \vdots 1]^T$, $\gamma^T = [\gamma_1 \ \cdots \ \gamma_n]$. The solution to this problem is obtained by the dual simplex method. The algorithm was illustrated by a 4 degree-of-freedom manipulator, moving in the three dimensional space, the obstacle being a tetrahedron (Fig. 6.9).

The above method for collision-free motion synthesis in redundant manipulators is the most complete, since it takes into account the distance between the obstacle and all the links, whereas the other methods provide only for the one or two points closest to the obstacle. It is aimed for off-line motion synthesis.

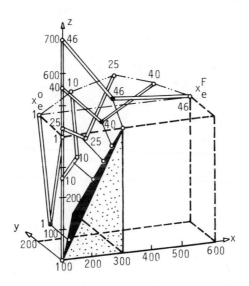

Fig. 6.9. An example of the optimal trajectory obtained by the linear programming method

An interesting method for redundant manipulator motion generation in the presence of obstacles and the constraints on joint coordinates, was proposed in [102]. The method provides for manipulator end-effector trajectory modification with the purpose of avoiding obstacles.

It is assumed that the manipulator hand trajectory is described by three Cartesian coordinates, $x_e = [x \ y \ z]^T$, being specified by

$$\dot{x}_e = f_x(x_e, t), \qquad x_e(0) = x_e^o = f(q^o), \qquad x_e(t) = f(q(t)) \qquad (6.4.27)$$

where $f_x(x_e, t): R^3 \times R \rightarrow R^3$ is a given function, $f(q): R^n \rightarrow R^3$ is the vector function relating between the joint coordinates and the manipulator hand position. In [102] the straight line motion according to

$$\dot{x}_e = f_x(x_e, t) = -\alpha(x_e - x_e^F) \qquad (6.4.28)$$

is considered, where x_e^F is the end point and α the scalar parameter defining the speed of approaching the end point x_e^F.

If the linearized kinematic model is considered, we have

$$\dot{x}_e = J(q)\dot{q} = f_x(x_e, t) \qquad (6.4.29)$$

where $J(q) = \partial f(q)/\partial q \in R^{3 \times n}$. In the general solution

$$\dot{q} = G\dot{x}_e + (I_n - GJ)y(t) \qquad (6.4.30)$$

the arbitrary vector $y(t)$ should be determined. Let us assume that the constraints on joint coordinates may be expressed in the form of a nonequality

$$\rho(q) > 0 \qquad (6.4.31)$$

where $\rho(q)$ is a scalar function describing the distance in the joint coordinates space between the state q and the "forbidden" set of states, defined by the obstacles or the physical limits on joint angles. In the case when $\rho(q) > \varepsilon$ is valid, the minimum norm solution is applied ($y(t) = =0$). However, when the nonequality

$$\rho(q(t)) \leqslant \varepsilon \qquad (6.4.32)$$

holds, with ε being a positive constant, the vector $y(t)$ is chosen from the condition $\dot{\rho}(q(t)) > 0$, i.e. from

$$\nabla\rho^T[G\dot{x}_e + (I_n - GJ)y] > 0 \qquad (6.4.33)$$

This nonequality is linear with respect to y. If the condition

$$(I_n - G\ J)\nabla\rho = 0 \tag{6.4.34}$$

is not fulfilled, it is possible to adopt y(t) in the form

$$y(t) = \frac{\lambda - \nabla\rho^T G\ \dot{x}_e}{||(I_n - G\ J)\nabla\rho||^2} (I_n - G\ J)\nabla\rho \tag{6.4.35}$$

where $\lambda > 0$. Thus the integration of (6.4.30) will ensure that the non-equality (6.4.31) will be still satisfied, i.e. that the manipulator will be far enough away from the constraint. It is possible also to introduce a set of nonequalities $\rho_1(q)$, $\rho_2(q)$,... of the type (6.4.31). In that case the procedure is carried out in an analogous manner.

If no solution to the nonequality (6.4.33) exists, the given manipulator tip trajectory $\dot{x}_e = f_x(x_e, t)$ cannot be realized, given the constraints (6.4.31). In this case it is necessary to modify the manipulator end-effector trajectory, so that the obstacle may be avoided. In [102] the proposed modification involves a temporary alteration of the end point on the straight line manipulator hand path.

Let us assume that at the time instant $t = t_1$ the nonequality (6.4.32) became true (i.e. that the manipulator reached the obstacle), while the Equation (6.4.34) holds (i.e. the selection of the y vector is not possible). Let the position of the manipulator tip at that time instant be $x_e^1 = f(q(t_1))$. In this case we will choose another trajectory end point x_e^2 from the following condition. By substituting the Equations (6.4.27) and (6.4.34) into the nonequality (6.4.33), it becomes

$$r^T(q(t_1))x_e^F - p(q(t_1)) > 0 \tag{6.4.36}$$

$$r(q) = G^T\nabla\rho, \qquad p(q) = r^T(q)f(q) \tag{6.4.37}$$

As we have said, the nonequality (6.4.36) cannot be satisfied. If $r(q) \neq 0$ holds, it is possible to replace the point x_e^F by x_e^2, which would satisfy $r^T(q(t_1))x_e^2 - p(q(t_1)) > 0$. There are many points which satisfy this condition. In [102] the point was chosen from the equation

$$x_e^2 = x_e^F + \frac{\delta + p(q(t_1)) - r^T(q(t_1))x_e^F}{||r(q(t_1))||^2} r(q(t_1)) \tag{6.4.38}$$

where δ is a given scalar value.

The motion synthesis is then carried out as if the point $x_e^1 = f(q(t_1))$ is the initial point and x_e^2 - the end point, according to (6.4.27), (6.4.29). The manipulator is moved along this segment of the straight line until the nonequality (6.4.36) is not fulfilled, i.e. until the obstacle prevents the direct movement from the instant position to the final point x_e^F. At the time the condition (6.4.36) becomes true, the end point x_e^2 is again replaced by the point x_e^F. Point replacement may be repeated several times if necessary.

An example of the trajectory of a 3 degree-of-freedom manipulator moving in a plane is shown in Fig. 6.10.

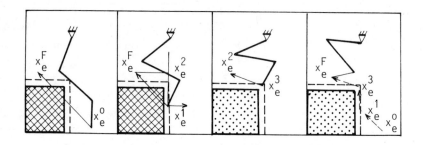

Fig. 6.10. The modification of the manipulator trajectory

The method presented provides for automatic trajectory modification. However, the problem arises of how to choose the parameters λ and δ automatically. The parameter λ should be chosen as a compromise between the values corresponding to a sufficiently fast motion away from the obstacle, and the values which would not cause high peaks in velocities $\dot{q}$ at the time instants when (6.4.32) becomes true. Similarly, the selection of the given parameter δ should ensure that the point x_e^2 does not drift too far away from the obstacle and the point x_e^F (e.g. in Fig. 6.10 it was chosen in such a way that the segment $x_e^1 - x_e^2$ is parallel to the obstacle surface).

The methods presented in this section represent the main results achieved so far in the domain of redundant manipulator motion synthesis in the presence of obstacles.

6.5. An Algorithm for Redundant Manipulator Motion Synthesis in the Presence of Obstacles

In this section we will present a real-time implementable algorithm for redundant manipulator motion synthesis in an obstacle-cluttered environment [105-106]. This procedure is based on the application of the performance indices which take the presence of the obstacles into account and thus prevent the arm from drawing too close to the obstacle.

This problem considered is the same as in the previous section. Nemely, the initial manipulator configuration $q^o \in Q \subset R^n$ is given, as well as the manipulator end-effector trajectory in the external coordinates space $x_e(t) \in R^m$, $t \in [0, T]$ between the initial point $x_e(0) = x_e^o = f(q^o)$ and the final point x_e^F. The disposition of obstacles is also given. The problem is to find the appropriate joint motion for the avoidance of obstacles.

As before, we will assume that the information on the distance between the critical point on the arm and the obstacle, and the location of the critical point x_c is available from a higher control level. The definition of the distance is introduced by Equations (6.4.1)-(6.4.3) in Section 6.4 (see Fig. 6.8).

In the case when the manipulator is far from the obstacle, i.e. $d(S, C) > > d_{max}$, the solution for the joint rates may be obtained in the usual manner, most frequently as the minimum norm solution, or by applying any of the procedures presented in Sections 6.2 and 6.3. However, when the distance between the critical point x_c on the arm and the obstacle lies between two given values

$$d_{max} > d(x_c, C) > d_{min} \qquad (6.5.1)$$

the presence of the obstacle should be taken into account, possibly by modifying the performance criteria so that motion is collission-free. We will here discuss the usefulness of several of the optimality criteria which may be applied.

The solution to the underdetermined linear system of equations

$$\dot{x}_e = J(q)\dot{q}, \qquad J \in R^{m \times n}, \qquad m < n \qquad (6.5.2)$$

is obtainable by applying the simple performance index in the joint coordinate space

$$\Omega(\dot{q}) = \frac{1}{2} \dot{q}^T M(q) \dot{q} \tag{6.5.3}$$

where $M(q) \in R^{n \times n}$ is a symmetric, positive definite matrix, given by

$$M(q) = \text{diag}(m_i), \qquad i=1,\ldots,n$$

$$m_i = \begin{cases} \dfrac{d_{max} - d_{min}}{d(x_c,C)-d_{min}}, & i=1,\ldots,\ell \\ \\ 1 & i=\ell+1,\ldots,n \end{cases} \tag{6.5.4}$$

Here ℓ denotes the number of the link to which the critical point belongs. The solution to the system (6.5.2) is then given by

$$\dot{q} = G \dot{x}_e, \qquad G = M^{-1} J^T (J M^{-1} J^T)^{-1} \tag{6.5.5}$$

This performance index "penalizes" the motion of those joints which influence the motion of the point x_c belonging to the ℓth link more heavily. Thus, the first ℓ joints are braked preventing the point x_c from approaching the obstacle. If the remaining $(n-\ell)$ joints are not sufficient to ensure the execution of the imposed task, a modification of the task is required.

The use of the index (6.5.3)-(6.5.4) is rather attractive because of its inherent simplicity.

A generalization of the performance index (6.5.3) is

$$\Omega(\dot{q}) = \frac{1}{2}(\dot{q}-\dot{q}_r)^T M(\dot{q}-\dot{q}_r) \tag{6.5.6}$$

where $\dot{q}_r \in R^n$ is a given reference velocity vector and $M \in R^{n \times n}$ is a symmetric positive definite matrix. In this case the index contains a linear term with respect to $\dot{q}$. It is equivalent to the criterion

$$\Omega(\dot{q}) = \frac{1}{2} \dot{q}^T M \dot{q} + M_1^T \dot{q} \tag{6.5.7}$$

where $M_1^T = -\dot{q}_r^T M$, which yields the optimal solution in the form

$$\dot{q} = G \dot{x}_e + (G J - I_n) M^{-1} M_1 \tag{6.5.8}$$

or

$$\dot{q} = G \, \dot{x}_e + (I_n - GJ) \dot{q}_r \tag{6.5.9}$$

where G is still given by (6.5.5). The problem of how to determine the reference joint velocities $\dot{q}_r$ automatically, so as to stay away from the obstacle is not simple, owing to the fact that the relationship between the joint coordinates and link positions is highly nonlinear and ambiguous for redundant manipulators. For this reason, we will propose a performance index in the external coordinates space in the sequel.

The modification of the index (6.5.3) is aimed at minimizing the linear velocity of the critical point x_c. Let us consider

$$\Omega(\dot{q}) = \frac{1}{2}\dot{q}^T M \, \dot{q} = \frac{1}{2}[\dot{q}^{LT} \ \dot{q}^{UT}] \begin{bmatrix} M^L & O_{\ell,n-\ell} \\ \hline O_{n-\ell} & I_{n-\ell} \end{bmatrix} \begin{bmatrix} \dot{q}^L \\ \hline \dot{q}^U \end{bmatrix} \tag{6.5.10}$$

where $\dot{q}^L = [\dot{q}_1 \ \cdots \ \dot{q}_\ell]^T$, $\dot{q}^U = [\dot{q}_{\ell+1} \ \cdots \ \dot{q}_n]^T$, ℓ - is the number of the link closest to the obstacle, $I_{n-\ell}$ - is $(n-\ell) \times (n-\ell)$ identity matrix, $O_{j,k} \in R^{j \times k}$ - the null matrix, $M^L \in R^{\ell \times \ell}$. The performance index (6.5.10) may be written as

$$\Omega(\dot{q}) = \Omega_1(\dot{q}^L) + \Omega_2(\dot{q}^U), \tag{6.5.11}$$

where

$$\Omega_1(\dot{q}^L) = \frac{1}{2} \dot{q}^{L^T} M^L \dot{q}^L \quad \text{and} \quad \Omega_2(\dot{q}^U) = \frac{1}{2} \dot{q}^{U^T} I_{n-\ell} \dot{q}^U \tag{6.5.12}$$

If it is to prevent the motion of the critical point x_c, the first part of the criterion becomes

$$\Omega_1(\dot{q}^L) = \frac{1}{2} \dot{x}_c^T M_c \dot{x}_c \tag{6.5.13}$$

where $\dot{x}_c = [\dot{x} \ \dot{y} \ \dot{z}]^T \in R^3$ is the linear velocity of the critical point expressed in the reference frame, $M_c \in R^{3 \times 3}$ - is the matrix of the performance index in the external coordinate space. The expression (6.5.13) can be transformed into the form (6.5.12) by introducing the relation

$$\dot{x}_c = J_c \dot{q}^L \tag{6.5.14}$$

into the Equation (6.5.13), where $J_c \in R^{3 \times \ell}$ is the Jacobian matrix for the point x_c. The matrix M^L then becomes

$$M^L = J_c^T M_c J_c \qquad (6.5.15)$$

The matrix M_c can be selected in different ways. Here, we suggest the form given by (6.5.4).

Accordingly, the motion of the first ℓ d.o.f. is optimized so as to ensure that the point x_c does not approach the obstacle. The motion of the remaining $(n-\ell)$ joints is governed by the standard criterion of minimal norm of the joint velocities vector.

However, it may also happen that, if the obstacle is close to some of the final links in the kinematic chain, the required task cannot be realized by stopping the critical point x_c, although this would eventually be possible if the point x_c moved. Hence, a generalization of the performance index $\Omega_1(\dot{q}^L)$ (6.5.13) is proposed

$$\Omega_1(\dot{q}^L) = \frac{1}{2}(\dot{x}_c - v_c)^T M_c (\dot{x}_c - v_c) \qquad (6.5.16)$$

where $v_c \in R^3$ is a given velocity vector of the point x_c, chosen in such a way that the point goes away from the obstacle. However, index (6.5.16) cannot be reduced to form (6.5.12), but to the form

$$\Omega_1(\dot{q}^L) = \frac{1}{2}\dot{q}^{L^T} M^L \dot{q}^L + M_1^{L^T} \dot{q}^L$$

$$M_1^L = -(v_c^T M_c J_c)^T \qquad (6.5.17)$$

and M^L is still given by (6.5.15). The constant term $\frac{1}{2}v_c^T M_c v_c$ from index (6.5.16) is not taken into account since it has no influence on the optimal solution. The total performance index $\Omega(\dot{q})$ now becomes

$$\Omega(\dot{q}) = \frac{1}{2}\dot{q}^T M \dot{q} + M_1^T \dot{q} \qquad (6.5.18)$$

where the matrix M is given by (6.5.10), and $M_1 = [M_1^{LT} \mid 0 \cdots 0]^T \in R^n$. The solution to system (6.5.2) which is optimal with respect to (6.5.18) is given by (6.5.8).

Evidently, a more complex criterion requires greater computational complexity. Comparing to the motion computation with performance index (6.5.6), the optimal motion calculation with index (6.5.16) requires the calculation of the Jacobian matrix J_c for the critical point. The introduction of this matrix into the motion synthesis algorithm, must obviously include the problems connected with the singularities of J_c, which do not exist in the configuration space performance indices (6.5.3), (6.5.6).

The problem of selecting the velocity vector v_c in (6.5.16) should also be resolved at the higher control level. The main reason for this is that here information about the disposition of obstacles is available. This information can be used profitably when selecting the vector v_c. For example, in Fig. 6.11 three different situations are presented: in (a) v_c is chosen to be orthogonal to the obstacle, in (b) v_c is orthogonal to the link, while in (c) it would be desirable to control not only the position of the critical point, but also the orientation of the link.

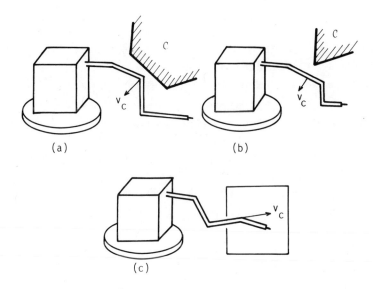

Fig. 6.11. Selection of the vector v_c

The performance index (6.5.16) offers excellent possibilities for obtaining a feasible trajectory, if such a solution physically exists. Whether a feasible trajectory will be completed or not, depends to a great extent on the way the velocity v_c is chosen at the higher control level.

Let us now discuss the numerical complexity of the proposed algorithm. Generally, this algorithm can be used both as off-line and on-line procedure for trajectory synthesis. Off-line synthesis would correspond to trajectory planning for a redundant manipulator in a known environment and for a given task. In that case the information on the distance between the arm and obstacles can be obtained by applying purely mathematical methods and obstacle modelling techniques, since trajectory calculation time is not critical.

If one is dealing with real-time application of the proposed algorithm, the information on the critical distance should be obtained from higher control level utilizing sensors. The proposed collision avoidance algorithm itself is convenient for real-time implementation, since it gives an explicit solution for joint rates at each sampling period, and employs no iterative procedures. Its computational complexity is modest compared with other proposed algorithms [100, 103]. It can be applied to moving objects, as well as to stationary obstacles.

The algorithm presented will be illustrated by an example of a robot with 5 revolute joints (Fig. 6.12). The first link rotates about a vertical axis, while the remaining joints have horizontal axes. The manipulation task to be performed is a straight line motion $x_e(t)$ of the manipulator tip $x_e(t) = x_e^O + \lambda(t)(x_e^F - x_e^O)$, $0 < t < T$ where $x_e = [x\ y\ z]^T$ are Cartesian coordinates of the manipulator end point with respect to the reference coordinate frame attached to the manipulator base; x_e^O and x_e^F are the initial and final points, respectively, and $\lambda[0, 1]$ is a scalar parameter which determines velocity distribution. In the examples to follow we shall adopt $\lambda(t) = (1-\cos(2\pi t/T))/T$, where T is the travelling time.

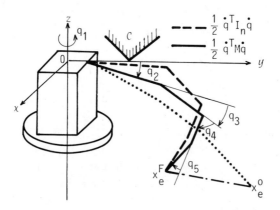

Fig. 6.12. Redundant manipulator motion - configuration
space performance index

In Figs. 6.12 and 6.13 motion in the vertical plane yOz between the points $x_e^O = [0. \ 0.91 \ -0.57]^T$[m] and $x_e^F = [0. \ 0.5 \ -0.5]^T$[m] is optimized. The initial configuration of the manipulator which corresponds to the point x_e^O is presented by dotted lines in these figures.

If the motion synthesis in the absence of obstacles, is performed applying the minimum norm of joint velocity vector, $(\Omega(q) = \frac{1}{2} \dot{q}^T I_n \dot{q})$, one obtains the manipulator configuration in the final point presented by broken lines in Figs. 6.12 and 6.13. On the other hand, if the synthesis is performed taking into account the presence of the obstacles in Figs. 6.12 and 6.13, using the performance index (6.5.3), the motion of the links is modified, so that the inboard links stay away from the obstacles. The parameters of the index are adopted to be $d_{max} = 0.1$ m, $d_{min} = 0.04$ m. The obtained configurations at the final point are presented by solid lines in these figures. It is evident that in these cases it was feasible to synthesize collision-free motion along the prescribed tip trajectory, i.e. that the imposed conditions were satisfied. Dynamic analysis, however, shows that the total energy consumption of actuators is increased (Fig. 6.14) due to the increased accelerations of some links. Therefore, the motion synthesis should also include a test as to whether the required task is realizable in the given time interval, i.e. whether the constraints on the maximal actuator input signals are satisfied.

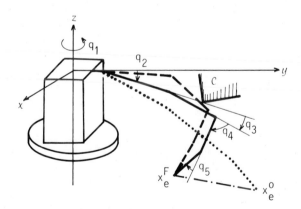

Fig. 6.13. Redundant manipulator motion - configuration space performance index

The illustration of the algorithm based on the performance index (6.5.16) is given by the example presented in Fig. 6.15. The manipulator described above moves along a vertical line between the points $x_e^O = [0. \ 0.63 \ -0.8]^T$ [m] and $x_e^F = [0. \ 0.63 \ - \ 0.5]^T$[m]. In this example the obstacle is closest

256

to the last but one link (4-th link). The initial configuration of the manipulator is presented by the dotted line in Fig. 6.15. The dashed line corresponds to the final configuration obtained in the absence of obstacle ($\Omega = \frac{1}{2}\dot{q}^T I_n \dot{q}$), while the solid line represents index (6.5.16). The velocity v_c was chosen to be orthogonal to the link. It is evident that the increased displacement of the second and third link enabled the manipulator to stay sufficiently far away from the obstacle.

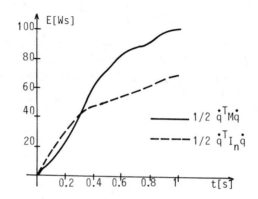

Fig. 6.14. Total energy consumption of the actuators

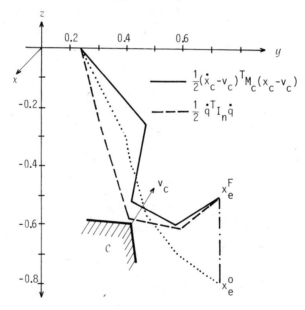

Fig. 6.15. Redundant manipulator motion - external coordinates performance index

Let us now discuss the problems that arise in redundant manipulator motion synthesis using indices (6.5.3), (6.5.6), (6.5.13), (6.5.16). The performance criteria in the external coordinates space involve the Jacobian matrix J_c of the critical point, so that elements of $M^L = J_c^T M_c J_c$ are frequently less than 1 (although the elements of M_c are greater than 1). In order to reconcile the numerical values of elements of M^L to the lower unit matrix $I_{n-\ell}$ (Eq. (6.5.10)), we have to normalize its elements by a constant. Here this constant was adopted to be the minimal diagonal element of $J_c^T J_c$. Problems also arise in the case when the critical point is far away from the manipulator base so that great differences exist between the Jacobian elements which describe the impact of the first and the critical joint on the motion of critical point x_c. This leads to singularity of M (6.5.13). In this case the degrees of freedom having a small impact on the manipulator hand motion should be eliminated from the weighting matrix. The same thing occurs in the case of the singularity of J_c. These problems do not appear with the configuration space performance indices.

A problem common to all these criteria arises if the braking zone $(d_{max} - d_{min})$ is rather narrow, and some of the links gain a high velocity before entering this zone. In this case, braking may be sudden which would lead to high acceleration and consequently high energy consumption. This may be partially avoided by extending the braking zone, although some prediction algorithms, based on the calculation of the critical distance gradient, could also be introduced.

Abrupt changes in the critical point closest to the obstacle may result in abrupt changes to the performance index matrix, and produce acceleration peaks. This problem arises also with all other algorithms proposed so far. Filtering these peaks would lead to deviation from the optimal solution, but would provide for the smoothness of motion.

If the obstacle is too close to the final link, so that the manipulation task cannot be realized, one should modify the manipulation task, taking deviation from the desired trajectory as a criterion and providing collision-free motion. This complex problem could be the subject of further research.

6.6. Summary

In this chapter we have discussed the problems connected with redundant manipulator motion synthesis. The various methods, aimed at resolving the ambiguity of the inverse kinematic problem, which have been proposed to date, have been presented. The motion synthesis in the free work space and in obstacle-cluttered environment, have been discussed separately. Various objectives may be achieved by employing the redundant manipulators: staying away from mechanical locks, obstacles, singularities, etc. Redundant manipulators will certainly play an important role in industrial robotics in the future.

References

[1] Paul R., Robots Manipulators: Mathematics, Programming, and Control, The MIT Press, 1981.

[2] Vukobratović M., Potkonjak V., Dynamics of Manipulation Robots: Theory and Application, Springer-Verlag, 1982.

[3] Popov E., Vereschagin A.F., Zenkevitch S.L., Manipulation robots: Dynamics and Algorithms, (in Russian), "Nauka", Moscow, 1978.

[4] Renaud M., "Calcul de la Matrice Jacobienne Nécessaire à la Commande Coordonée d'un Manipulateur", Mechanism and Machine Theory, Vol. 15, 1980.

[5] Vukobratović M., Stepanenko Y., "Mathematical Models of General Anthropomorphic Systems", Mathemat. Biosciences, Vol. 17, 1973, pp. 191-242.

[6] Stepanenko Y. and Vukobratović M., "Dynamics of Articulated Open-Chain Active Mechanisms", Math. Biosc. Vol. 28, 1976, pp. 137-170.

[7] Vukobratović M., Stokić D., Control of Manipulation Robots, Series: Scientific Fundamentals of Robotics 2, Springer-Verlag, 1982.

[8] Denavit J., Hartenberg R.S., " Kinematic Notation for Lower-Pair Mechanisms Based on Matrices", ASME J. of Applied Mechanics, June 1955, pp. 215-221.

[9] Whitney D.E., "The Mathematics of Coordinated Control of Prosthetic Arms and Manipulators", Trans. of the ASME, Jour. of Dyn. Syst., Meas. and Control, Vol. 94, No. 4, 1972.

[10] Vukobratović M., Kirćanski M., "Correlation Between Denavit-Hartenberg Manipulator Kinematics and Rodrigues Formula Approach", submitted for publication to Mechanism and Machine Theory.

[11] Abadie L., "Utilisation des Quaternions pour la Représentation des Coordonnées Angularies", Technique de Calcul Analogique et Numérique en Aéronautique, Congres de Liège, 1963.

[12] Segercrantz J., "New Parameters for Rotations of Solid Bodies", Commentatione Physico - Mathematicae, Societas Scientiarum Fennica, 33(2), 1966.

[13] Makino H., "A Kinematical Classification of Robot Manipulators", 6th Int. Symp. on Industrial Robots, Paper F2, 1976.

[14] Orin D., Schrader W., "Efficient Jacobian Determination for Robot Manipulators", VI IFToMM Congress, New Delhi, 1983.

[15] Pradin B., Contribution à l´étude de la stabilisation et de la commande de l´attitude de satellites an moyen de jets de gaz, Ph. Thesis, Toulouse, 1971.

[16] Vukobratović M., Kirćanski N., Real-Time Dynamics of Manipulation Robots, Series: Scientific Fundamentals of Robotics 4, Springer-Verlag, 1985.

[17] Vukobratović M., Kirćanski N., "Computer Assisted Generation of Robot Dynamic Models in Analytical Form", Acta Applicandae Mathematicae, International Journal of Applying Mathematics and Mathematical Applications, Vol. 2, No. 2, 1984.

[18] Vukobratović M., Kirćanski N., "Computer-Aided Procedure of Forming Robot Motion Equations in Analytical Forms", Proc. of VI IFToMM Congress on Theory of Mach. and Mech., New Delhi, 1983.

[19] Vereschagin A.F., Generozov V.L., "Planing of Manipulator Trajectories", (in Russian), Teknicheskaya Kibernetika AN USSR, No.2, 1978.

[20] Paul R., Shimano B., Mayer G., "Kinematic Control Equations for Simple Manipulators", IEEE Trans. on Systems, Man and Cyber., Vol. SMC-11, No. 6, June 1981.

[21] Paul R., Shimano B., Mayer G., "Differential Kinematic Control Equations for Simple Manipulators", IEEE Trans. on Systems, Man and Cyber., Vol. SMC-11, No. 6, June 1981

[22] Featherstone R., "Position and Velocity Transformations Between Robot End-Effector Coordinates and Joint Angles", International Journal of Robotics Research, Vol. 2, No. 2, 1983.

[23] Pieper D.L., Roth B., "The Kinematics of Manipulators Under Computer Control", Proc. II Inter. Congress on Theory of Machines and Mechanisms, Vol. 2, Zakopane, 1969.

[24] Pieper D.L., The Kinematics of Manipulators Under Computer Control, Ph. D. Thesis, Stanford University, Dept. of Mech. Eng., 1968.

[25] Whitney D.E.,"Resolved Motion Rate Control of Manipulators and Human Prostheses", IEEE Trans. on Man-Machine Systems, Vol. MMS-10, No. 2, June 1969.

[26] Whitney D.E., "Optimum Step Size Control for Newton-Raphson Solution of Nonlinear Vector Equation", IEEE Trans. Aut. Control, Vol. AC-14, No. 5, Oct. 1969.

[27] Vereschagin A.F., Generozov V.L., Kutcherov V.B., "Control Algorithms of Manipulator via Velocity Vector", (in Russian), Teknicheskaya Kibernetika, No. 3, 1975.

[28] Vereschagin A.F., Minaev L.N., "The Design Principles of Remote Position Controllers for Manipulation Robots", Teknicheskaya Kibernetika, No. 4, 1978.

[29] Ignatiev M.B., Kulakov F.M., Pokrovskij A.M., "Manipulation Robots Control Algorithms", Mashinostroenie, 1972.

[30] Malshev V.A., Timofeev A.V., "An Algorithm for Manipulator Trajectory Synthesis in the Presence of Obstacles and Mechanical Constraints", Teknicheskaya Kibernetika, No. 6, 1978.

[31] Malshev V.A., "A Method of Generating Programmed Manipulator Trajectories", Teknicheskaya Kibernetika, No. 6, 1980.

[32] Featherstone R., "Calculation of Robot Joint Rates and Actuator Torques from End Effector Velocities and Applied Forces", Mechanism and Machine Theory, Vol. 18, No. 3, 1983.

[33] Kobrinski A.A., Kobrinski A.E., "A Contribution to Manipulation System Motion Synthesis", Reports of USSR Academy of Sciences, No. 5, 1975.

[34] Renaud M., Contribution à la Modelisation et à la Commande Dynamique des Robots Manipulateurs, Ph. D. Thesis, Toulouse, 1980.

[35] Carnahan B., Luther H.A., Wilkes, "Applied Numerical Methods", John Willey & Sons, 1969.

[36] Dongarra J.J., Moler C.B., Bunch J.R., Stewart G.W., Linpack User's Guide, Siam, Philadelphia, 1979.

[37] Mutjaba Sh., Goldman R., AL User's Manual, 3rd ed., Computer Science Department, Stanford Univ., Palo Alto, CA, Rep. no. STAN-CS-81-889, Dec. 1981.

[38] IBM Robot System/1: AML Reference Manual, 2nd ed., IBM Corporation, Boca Raton, FL, SC34-0410-1, Sept. 1981.

[39] Taylor R.H., Summers P.D., Meyer J.M., "AML: A Manufacturing Language", Inter. Journ. of Robotics Research, Vol. 1, No. 3., pp. 19-41.

[40] Allegro Operator's Manual (A12 Assembly Robot), General Electric Co., Orlando, FL, Feb. 1982.

[41] Craig J.J., JARS: JPL Autonomous Robot System, Robotics and Teleoperators Group, Jet Propulsion Laboratory, Pasadena, CA, 1980.

[42] User's Guide to VAL, Version 12, Unimation, Inc., Danbury, CT, 398-H2A, June 1980.

[43] User's Guide to VAL-II, Part 1: Control from the System Terminal, Part 2: Communications with a Supervisory System, Part 3: Real-Time Path Control, version X2, Unimation, Inc., Danbury, CT, Apr. 1983.

[44] Bonner S., Shin K., "A Comparative Study of Robot Languages", Computer, December 1982, pp. 82-96.

[45] Gruver W., Soroka B., Craig J., Turner T., "Industrial Robot Programming Languages: A Comparative Evaluation", IEEE Trans. on Syst., Man, and Cybernetics, Vol. SMC-14, No. 4, July/August 1984.

[46] Wood B.O., Fugelso M.A., "MCL, The Manufacturing Control Language", 13th ISIR, Chicago, April 1983, pp. 12-84-12-96.

[47] Paul R., et al, "Advanced Industrial Robot Control Systems", First Report, Purdue University, 1978.

[48] Luh J.Y.S., Walker M.W., "Minimum-time Along the Path for a Mechanical Arm", Proc. of the IEEE Conference on Decision and Control, Vol. 1, New Orleans, 1977, pp. 755-759.

[49] Luh J.Y.S., Lin C.S., "Optimum Path Planning for Mechanical Manipulators", Trans. of the ASME, Journal of Dyn. Syst., Meas. and Control, Vol. 102, No. 2, June 1981, pp. 142-152.

[50] Takase K., Paul R., Berg J., "A Structured Approach to Robot Programming and Teaching", IEEE Trans. on Syst., Man and Cybernetics, Vol. SMC-11, No. 4, April 1981.

[51] Paul R.P., Manipulator Cartesian Path Control, IEEE Trans. on Systems, Man, and Cybernetics, SMC-9, No. 11, Nov. 1979, pp. 702--711.

[52] Paul R., Zhang Hong, "Robot Motion Trajectory Specification and Generation", II Inter. Symp. on Robotics Research, Kyoto, 1984, pp. 373-380.

[53] Udupa S.M., "Collision Detection and Avoidance in Computer Controlled Manipulators", Proc. 5th Int. Conf. Artificial Intelligence, MIT, Cambridge, MA, Aug. 1977.

[54] Lozano-Pérez T., "Automatic Planning of Manipulator Transfer Movements", IEEE Trans., Syst., Man, and Cybern., Vol. SMC-11, No. 10, Oct. 1981, pp. 681-698.

[55] Lozano-Pérez T., "Spatial Planning: A Configuration Space Approach", IEEE Trans. on Computers, Vol. C-32, 1983, pp. 108-120.

[56] Brooks R.A., "Solving the Find-path Problem by Good Representation of Free Space", IEEE Trans. on Systems, Man and Cybern., Vol. SMC--13, No. 3, March/April 1983, pp. 190-197.

[57] Castain R., Paul R., "An On-line Dynamic Trajectory Generator", The Inter. Journal of Robotics Research, Vo. 3, No. 1, 1984, pp. 68-72.

[58] Vukobratović M., Stokić D., Kirćanski N., Nonadaptive and Adaptive Control of Manipulation Robots, Series: Scientific-Fundamentals of Robotics 5, Springer-Verlag, 1985.

[59] Kahn M.E., Roth B., "The Near-Minimum-Time Control of Open-Loop Articulated Kinematic Chains", Trans. of the ASME, Journ. of Dynamic Syst., Meas., and Control, Sept. 1971, pp. 164-172.

[60] Bolotnik N.N., Kaplunov A.A., "Optimal Straight-Line End-Effector Motion of a Two-Link Manipulator", Teknicheskaya Kibernetika, No. 1, 1982.

[61] Cvetković V., Vukobratović M., "Contribution to Controlling Non--Redundant Manipulators", Mechanism and Machine Theory, Vol. 16, No. 1, 1980, pp. 81-91.

[62] Heimann B., Losse H., Schuster G., "Contribution to Optimal Control of an Industrial Robot", IV CISM-IFToMM Simposium on Theory and Practice of Robots and Manipulators, Zaborów near Warsaw, 1981.

[63] Dubowsky S., Shiller Z., "Optimal Dynamic Trajectories for Robotic Manipulators", V-th CISM - IFToMM Symp. on Theory and Practice of Robots and Manipulators, Udine, Italy, June 1984.

[64] Shih L.Y., "On the Elliptic Path of an End-Effector for an Anthropomorphic Manipulator", The International Journal of Robotics Research, Vol. 3, No. 1, Spring 84, pp. 51-57.

[65] Hollerbach J.M., "A Recursive Formulation of Lagrangian Manipulator Dynamics", IEEE Trans. on Systems, Man and Cybernetics, Vol. SMC-10, No. 11, 1980, pp. 730-736.

[66] Luh J.Y.S., Walker M.W., Paul R.P.C., "On-line Computational Scheme for Mechanical Manipulators", Trans. of ASME Journal of Dynamic Systems, Measurement and Control, Vol. 102, No. 2, 1980, pp. 69--76.

[67] Vukobratović M., Kirćanski M., "One Method for simplified Manipulator Model Construction and its Application in Quazioptimal Trajectory Synthesis", Mechanism and Machine Theory, Vol. 17, No. 6, 1982, pp. 369-378.

[68] Cvetković V., Vukobratović M., "Computer-Oriented Algorithm of Variable Complexity for Mathematical Modelling of Active Mechanisms", IEEE Trans. on Systems, Man and Cybernetics, Vol. SMC-12, No. 6, 1982, pp. 838-848.

[69] Kirćanski N., Contribution to Dynamic Modelling and Control of Manipulation Robots, Ph. D. Dissertation, Faculty of Electrical Engineering, Belgrade University, 1984.

[70] Merrit E.H., Hydraulic Control Systems, John Willy & Sons, 1967.

[71] Borovac B., Vukobratović M., Stokić D., "Analysis of the Influence of Actuator Model Complexity to Manipulator Control Synthesis", Mechanism and Machine Theory, Vol. 18, No. 2, 1983.

[72] Athans M., Falb P.L., Optimal Control: An Introduction to the Theory and Its Application, McGraw Hill, New York, 1966.

[73] Bryson A., Yu-Chi Ho, Applied Optimal Control: Optimization, Estimation and Control, John Wiley & Sons, New York, 1975.

[74] Vukobratović M., Kirćanski N., "Computer-Oriented Method for Linearization of Dynamic Models of Active Spatial Mechanisms", Mechanism and Machine Theory, Vol. 17, No. 1, 1982, pp. 21-32.

[75] Kirćanski M., Contribution to Manipulation Robot Control Synthesis, M.Sc. Thesis, Faculty of Electrical Engineering, Belgrade University, 1980.

[76] Vukobratović M., Kirćanski M., "A Method for Optimal Synthesis of Manipulation Robot Trajectories", Trans. of ASME, Jour. of Dynamic Systems, Measurement and Control, Vol. 104, No. 2, June 1982, pp. 188-193.

[77] Kirćanski M., "Contribution to Synthesis of Nominal Trajectories of Manipulators via Dynamic Programming", I International Conference

on System Engineering, Coventry, G.Britain, 1980.

[78] Bellman R., Dynamic Programming, Princeton University Press, Princeton, N.J., 1957.

[79] Vukobratović M., Stokić D., Kirćanski M., "Contribution to Dynamic Control of Industrial Manipulators", XI International Symposium on Industrial Robots, Tokyo, 1981.

[80] Vukobratović M., Cvetković V., Kirćanski M., Kinematic and Dynamic Approach to Motion Synthesis for Manipulation Robots, "Mihailo Pupin" Institute Press, Belgrade, 1983.

[81] Liégois A., "Automatic Supervisory Control of the Configuration and Behaviour of Multibody Mechanisms", IEEE Trans., Syst., Man and Cybernetics, Vol. SMC-7, No. 12, 1977, pp. 868-871.

[82] Fournier A., Khalil W., "Coordination and Reconfiguration of Mechanical Redundant Systems", Proc. Int. Conf. on Cybernetics and Society, Washington, D.C., 1977, pp. 227-231.

[83] Renaud M., Contribution a L'étude de la Modélisation et de la Commande des Systèms Mécaniques Ariticulés, Ph.D. Thesis, Université Paul Sabatier de Toulouse, 1975.

[84] Kobrinski A.A., Kobrinski A.E., "Optimal Motion Synthesis for Manipulation Systems", Mashinostroenie, No. 1, 1976.

[85] Kozlov J.M., Mayorov A.P., "A System for Programmed Robot Control", Teknicheskaya Kibernetika, No. 5, 1976.

[86] Ignatiev M.B., et al., "On Algorithms of Computer Controlled Manipulators", Advances in External Control of Human Extremities - ETAN Publication, Belgrade, 1970, pp. 365-394.

[87] Konstantinov M.S., Patarinski S.P., "A Contribution to the Inverse Kinematic Problem for Industrial Robots", 12-th ISIR, Paris, 1982, pp. 459-467.

[88] Klein Ch.A., Ching-Hsian Huang, "Review of Pseudoinverse Control for Use with Kinematically Redundant Manipulators", IEEE Trans. on Syst., Man, and Cyber., Vol. SMC-13, No. 3, March/April 1983.

[89] Yoshikawa T., "Analysis and Control of Robot Manipulators with Redundancy", Ist Intern. Symp. of Robotics Research, Bretton Woods, USA, 1983.

[90] Searle S.R., Linear Models, New York: Wiley, 1971.

[91] Boullion T.L., Odell P.L., Generalized Inverse Matrices, New York, Wiley Interscience, 1971.

[92] Lovass-Nagy V., Miller R., Powers D., "An Introduction to the Application of the Simplest Matrix-Generalized Inverse in Systems Science", IEEE Trans. on Circuits and Systems, Vol. CAS-25, No. 9, September 1978, pp. 766-771.

[93] Penrose R., "A Generalized Inverse for Matrices", Proc. Combridge Phil, Soc., Vol. 51, 1981, pp. 406-413.

[94] Albert A., Regression and the Moore-Penrose Pseudoinverse, Acad. Press, New York - London, 1972.

[95] Klema V.C., Laub A.J., "The Singular Value Decomposition: Its Computation and Some Applications", IEEE Trans. Autom. Contr., Vol. AC-25, No. 2, Apr. 1980.

[96] Garbow B.S. et al., Matrix Eigensystem Routines - EISPACK Guide Extension, Springer-Verlag, Heidelberg, New York, 1977.

[97] Vukobratović M., Kirćanski M., "A Dynamic Approach to Nominal Trajectory Synthesis for Redundant Manipulators", IEEE Trans. on Systems, Man, and Cybernetics, Vol. 14, No. 4, 1984.

[98] Vukobratović M., Kirćanski M., "New Method for Nominal Trajectory Synthesis for Redundant Manipulators", Journal of USSR Academy of Sciences, Mashinovedenie, No. 4, 1984, pp. 21-26.

[99] Kobrinski A.A., Kobrinski A.E., "Manipulation System Trajectory Synthesis in the Environment with Obstacles", Comm. of Academy of Sciences of USSR, T. 224, No. 6, 1975.

[100] Khatib O., Le Maitre F.J., "Dynamic Control of Manipulators Operating in a Complex Environment", Proc. 3rd Int. CISM-IFToMM - Symp., Udine, Sept. 1978, pp. 267-282.

[101] Generozov V.L., "Algorithms for Trajectory Planning in the Presence of Obstacles", Teknicheskaya Kibernetika, No. 1, 1984.

[102] Aksenov G.S., Voroneckaya D.K., Fomin V.N., "Computer-Aided Planning of Programmed Manipulator Motions", Teknicheskaya Kibernetika, No. 4, 1978, pp. 50-55.

[103] Klein A. Ch., "Use of Redundancy in the Design of Robotic Systems", II International Symposium on Robotics Research, Kyoto, 1984.

[104] Rockafellar R.T., Convex Analysis, Princeton University Press, Princeton, New Jersey, 1970.

[105] Kirćanski M., Vukobratović M., "Contribution to Control of Redundant Robotic Manipulators in Environment with Obstacles", Intern. Journal of Robotics Research, to appear in 1985.

[106] Kirćanski M., Vukobratović M., "Trajectory Planning for Redundant Manipulators in the Presence of Obstacles", V CISM-IFToMM Symposium on Theory and Practice of Robots and Manipulators, Udine, 1984.

[107] Takano M., Yashima K., Toyama S., "A New Method of Solution of Synthesis Problem of a Robot and its Application to Computer Simulation System", 14-th International Symposium on Industrial Robots, Gothenburg, Sweden, 1984.

[108] Chun-Shin Lin, Po-Rong Chang, "Joint Trajectories of Mechanical Manipulators for Cartesian Path Approximation", IEEE Trans. on Systems, Man and Cybernetics, Vol. SMC-13, No. 6, Nov./Dec. 1983.

Subject Index